Financing Agriculture
into the Twenty-first Century

Financing Agriculture into the Twenty-first Century

EDITED BY

Marvin Duncan and
Jerome M. Stam

Routledge
Taylor & Francis Grou

LONDON AND NEW YORK

First published 1998 by Westview Press

Published 2019 by Routledge
52 Vanderbilt Avenue, New York, NY 10017
2 Park Square, Milton Park, Abingdon, Oxon OX14 4RN

Routledge is an imprint of the Taylor & Francis Group, an informa business

Library of Congress Cataloging-in-Publication Data
Financing agriculture into the twenty-first century / edited by Marvin
    Duncan, Jerome M. Stam.
        p. cm.
    Includes bibliographical references.
    1. Agriculture—Finance—Forecasting. 2. Agricultural credit—
Forecasting. 3. International finance—Forecasting.
4. International economic relations—Forecasting. 5. Agriculture—
United States—Finance—Forecasting. 6. Agricultural credit—
United States—Forecasting. 7. Twenty—first century—Forecasts.
I. Duncan, Marvin R. IT. Stam, Jerome M.
HD1437.F56 1998
332.7'1-dc21   98-22831
                                                          CIP

ISBN 13: 978-0-367-00988-5 (hbk)

Contents

Tables, Figures, and Boxes

Boxes

Foreword

This book explores the relationship between agriculture and the financial markets and challenges us to look ahead to the twenty-first century and answer the questions: Who will offer credit to agriculture? To whom will they make credit available? And, how will the credit be provided? The needs of agricultural firms for financial services are different now than they were in the past. And, we can expect that in the future, these needs will change again. The structure of U.S. agriculture (firm size, organization, and ways of doing business) will continue to change and evolve, resulting in new financing needs, new products, and new lending policies and procedures.

U.S. agriculture is becoming more industrialized and the private and public providers of credit are undergoing significant restructuring. As these changes converge, new relationships will evolve that portend the answers to the questions—Who will offer credit? To whom? And how? These questions challenged several leading specialists in agricultural economics and agricultural finance to develop a comprehensive analysis of the current trends affecting agriculture, the changes occurring in the financial sector, and the likely outcomes of their intersection. That analysis constitutes this book.

The future structure of U.S. agriculture is being shaped by many forces, some of which we cannot control, others that we can and will try to control, and still others that we could and will decide not to control. Many forces are driving change:

- General economic forces associated with the globalization of the economy.
- New technology.
- Changing attitudes about the use of natural resources and the environment.
- Increasing numbers of potential consumers and changing tastes of consumers.
- The natural forces associated with weather and the possibility of climate change.
- The availability and cost of credit, i.e., the financial environment within which agriculture operates.

Thus, we see the interaction of these and other forces. Not only will the structure of agriculture influence the demand for financial services, but the financial market changes will influence the future structure of agriculture by determining who will receive credit and how. For example, as contractual arrangements become more common across agriculture, will lenders lend to the producers or to the contractors? Will larger agricultural production units be favored over smaller operations? What conditions will lenders impose to limit their risk?

As we look ahead, we must remember the lessons of the past. The 1970s and 1980s taught us a powerful lesson about the relationship between agricultural and financial markets. With an expanding money supply and easy credit, agriculture boomed during the 1970s with an undervalued dollar boosting export demand, relatively low interest rates, and inflating land values. Agriculture was on a joy ride until the early 1980s, when the Federal Reserve Board hit the brakes beginning in October 1979 and U.S. agriculture experienced whiplash. With a restricted money supply, higher interest rates, and increased value of the dollar, export demand decreased, land values dropped, and many agricultural producers were left facing bankruptcy.

Moreover, this book has a broad focus ranging from the changing structure of agriculture to the growing category of nontraditional lenders (captive finance companies). Financing agriculture involves not only a wide variety of lenders and farm firms, but in a broader sense also the related input suppliers, marketers, and processors. The audience of this book therefore will be broad and varied in scope, ranging among academics, students, lenders, and agribusiness leaders. The information in this book is needed by progressive agricultural producers as they position their operations for the twenty-first century. Likewise, agribusiness firms, food processors, and educators and researchers need this longer term perspective for program planning and priority setting.

The authors of this book present their visions of the future based on their analyses of recent history and current trends. But because the future is uncertain, this book cannot be the last word on agriculture and its financing needs. Its primary role may be to provide a framework to promote further discussion about these future changes and their implications for decisionmakers in agriculture, in the financial services sector, and in the public sector. By stimulating further reflection and study based on a common understanding of the current trends and possible implications, the book is useful in helping the agricultural finance sector prepare itself for the future.

We are facing many new challenges in agriculture and agricultural finance with the evolution of new relationships between agriculture and the financial sector. We need to understand these new relationships, and that is where this book contributes by analyzing the current state of affairs and what is likely to evolve from the changes underway in the structure of agriculture, the institutions lending to agriculture, and in the methods for carrying out this lending function. We are reminded that many variables and several interrelated factors will shape the final outcomes and the future of U.S. agriculture.

A. Gene Nelson
Professor and Head
Department of Agricultural Economics
Texas A&M University
College Station, Texas

Preface

Eighteen years have passed since a major, industrywide symposium resulted in the publication of the book *Future Sources of Loanable Funds for Agricultural Banks* by the Federal Reserve Bank of Kansas City. Twelve years have elapsed since the publication by Westview Press of the comprehensive book *Financing the Agricultural Sector: Future Challenges and Policy Alternatives*, written by Dean W. Hughes, Stephen C. Gabriel, Peter J. Barry, and Michael D. Boehlje. Agricultural sector and agricultural lender conditions have changed dramatically since these books were published. The first was completed prior to the farm financial crisis of the early to mid-1980s, while the latter appeared at the end of this unusual period.

There is thus a need for a new book that is broadly focused on issues of structural change in agriculture and on the lenders serving its demand for credit. Since the early to mid-1980s, changes in technology and in business strategies have resulted in a quickening of the pace of structural change in both production agriculture and agricultural lending. Moreover, the economic stakes associated with those changes and the resultant demands placed on those who lend to agriculture have risen.

This book addresses those changes and focuses on the implications for those who seek to fill the credit demand of agriculture. It seems apparent to us that even greater changes in the delivery of agricultural financing are in the offing. Thus, we believe this book will be of value both to those who study the financing of a changing agriculture and those whose business is to provide that financing.

Many persons deserve appreciation for their efforts in influencing and producing this book. We particularly are indebted to the book's chapter authors for their thoughtful analyses that comprise this book. A number others have enhanced the quality of the final product and we can only highlight a few. The early critique of this project by Michael D. Boehlje, Laurence M. Crane, and Cole R. Gustafson is appreciated. We are especially indebted to the speakers and other participants in the American Agricultural Economics Association Pre-Conference Workshop "Financing North American Agriculture into the Twenty-First Century," held on July 27, 1997 at Toronto, Ontario, for numerous helpful insights for both the editors and authors. We are indebted to Dale B. Simms for his editorial efforts in helping to put the book's chapters into camera-ready form. Finally, we are most gratified that Westview Press has chosen to publish this book.

Marvin Duncan
Jerome M. Stam

Introduction

Marvin Duncan and Jerome M. Stam

The business and competitive environment within which firms lending to farmers and agribusinesses operate will be fundamentally changed in the years ahead. Technological advances are changing the way farmers produce their products. New business linkages across farms, input suppliers, marketers, and processors are changing the way farmers organize their businesses. This book identifies and explains the major challenges and opportunities confronting those firms that lend to agricultural producers and agribusinesses as they prepare for the twenty-first century.

Government-sponsored lending to the agricultural sector is changing as well. The Congress and executive branch have taken steps to reduce the role of direct government lending to farmers through cutbacks in the funding for programs of the Farm Service Agency (FSA). Direct lending programs that remain will be focused increasingly on narrowly defined client groups, and be required to meet more stringent tests of added public value. It is, however, unclear just how deep cuts will be in programs to guarantee lending to otherwise uncreditworthy farmers. Though Congress has not taken steps to fundamentally broaden government-sponsored enterprise (GSE) lending activity—such as increasing the authority of the Farm Credit System (FCS) and Farmer Mac—it has over the past several years provided incremental increases in GSE authorities. Moreover, government lending programs focused on rural economic development and export financing continue to be funded.

Farms are changing, growing larger across the range of commercial-scale farms, developing more complex business relationships among suppliers, marketers, and processors, and becoming technologically more advanced. What farms are not doing, however, is evolving into two easily identifiable and categorized segments based on size. Instead, while the trends toward scale, complexity, and technological advancement are pervasive across commercial-scale farms, smaller, specialized, or simpler business enterprises remain abundant and offer interesting market niches to lenders who wish to concentrate on certain market segments. What is obvious is that one size no longer fits all, and the diverse credit needs of the farm sector will increasingly require market segmentation and specialization of financial products suited to the needs of customers in the targeted market segment.

Lenders and other providers of financial services to the food and fiber sector are undergoing change as well. As lenders seek to achieve greater efficiencies and market shares, they are focusing on serving certain cohorts of farmers and agribusinesses, and not serving others. As a result, some farmers may find a great deal of competition for their business, while other farmers may have to search for a lender interested in providing the services they require. Some farmers will enjoy a personalized relationship with the same lender over a long period of time. Other farmers will make do with a series of transaction relationships with lenders, the duration of which may not extend beyond a single credit request.

The distinction between lenders and other financial services providers will blur as statutory and regulatory barriers come down and nationwide branch banking becomes a reality. Some lenders will seek to become a financial services superstore, while other lenders will cultivate a carefully defined market, seeking to serve the full needs of that market. Still others will become narrowly defined specialists with few products, but great capacity to deliver them.

The lending and financial services field will become much more crowded as most larger supply, marketing, and processing firms doing business with farmers seek to develop credit or other financial services products to market. That competition will become more international in character as ownership of firms doing business with farmers and agribusinesses becomes more transnational. That will present new challenges for banking regulators and for both public and private policymakers.

Evaluating credit proposals will become more complex. A host of alliances, contractual arrangements, joint ventures, and interlocking ownership arrangements across different segments of the food and fiber industry will proliferate as the agricultural sector becomes larger and more complex. Sorting out where costs are incurred and profits are earned, and evaluating creditworthiness that increasingly will be premised on the strength of proprietary specialized knowledge, technology, and business relationships will challenge lenders and borrowers alike.

Because of these fundamental changes, we have chosen to organize this book around issues, relationships, and structural changes, rather than simply a review of the impact of changes on separate lending groups, such as commercial banks or the FCS. Because of the magnitude of the agricultural industry transition now under way, we believe the traditional approach of analyzing finance and credit issues within a narrow context to be inadequate. We believe the book's focus on the broader issues, relationships, and structural changes and the effects of those on the agricultural sector will prove engaging and useful to lenders, borrowers, and students alike.

PART ONE

FORCES INDUCING CHANGE

1

Macroeconomic Factors and International Linkages Affecting the Financing of Agriculture in a World Economy

Glenn D. Pederson, Jeffrey T. Stensland
and Martin L Fischer

The implications of macroeconomics for the agricultural sector have received greater attention since the late 1970s, because of the increasing dependence of the sector on national and international economic/financial forces. Increasingly, those forces are communicating greater volatility to the agricultural sector via financial and commodity markets. With the increase in volatility and stronger linkages between those markets, there is also a greater need for agricultural producers and their lending institutions to manage the associated risks in more innovative and effective ways.

Changing National and Global Financial Markets

The lack of previous attention to macroeconomic issues such as monetary and fiscal policy, exchange rate regimes, international capital markets. and international trade policy has been attributed to three factors (Schuh 1974). First, greater attention was given to the sectoral approach and less to the national and international dimensions. Second, national economies were frequently treated as closed, so that exchange rates and capital flows were largely ignored. Third, exchange rates were fixed during the post-World War II, pre-Bretton Woods era. These conditions have been altered by a number of trends that are expected to continue into the twenty-first century.

The Framework of an Open Economy

To explore issues pertinent to the international dimensions of agricultural finance and trade, it is necessary to look at macroeconomic events in the context of an open economy. An open economy is characterized by the flow of capital and commodities into and out of the various national economies in response to changing interest rates and prices. These flows integrate national economies into a set of interrelated global markets for capital, currencies, commodities, and related factors of production. Smoothly functioning international financial markets are an important element of that global market system. By reducing the costs of capital transactions, it is possible for international financial markets to create increased economic efficiency and to expand the volume of trade.

Keynesian macroeconomic theory serves as a useful starting point to consider the concept of a closed versus an open economy. In *The General Theory of Employment, Interest, and Money*, there is no mention of exchange rates, international aspects of income determination, or the role of international efforts to create national stability (Keynes 1936). The Keynesian theory focused on the national economy and, because of that focus, the prices of domestic goods, the prices of domestic assets, and interest rates were determined in domestic markets. Meade (1951) later adapted the Keynesian analysis to international economic problems. However, this and other simplified "open economy" models neglected the link between domestic and foreign interest rates. A significant change occurred later in the development of these open economy models as the introduction of balance-of-payments analysis shifted the focus from goods markets to money and bond markets and onto the important role of interest rates.

National economies were more insular following World War II and closer to the simple open economy model in three respects: (1) the share of foreign trade transactions was relatively small, and significant trade barriers existed so that exchange rates could be changed without affecting domestic prices; (2) the international capital market did not function freely, so deficits and surpluses were not as easily financed with private capital flows; and (3) national monetary systems were insulated so that interventions in foreign exchange markets did not affect the monetary base (i.e., the effects of intervention were sterilized) and interest rates could be effectively controlled by monetary policy (Kenen 1985, p. 636). With the breakdown of pegged exchange rate schemes in the post-Bretton Woods era, there was an increase in capital mobility and, with it, an opening of the insular economies.

"Openness" is an average concept, since it does not clearly measure the extent to which foreign transactions affect domestic markets. Grassman (1980) suggests the trade-to-income ratio as an indicator of the openness of an economy. An increase in the trade-to-income ratio implies greater international linkages between the economy (or sector) and the rest of the world due to an increase in trade activity. The average trade-to-income percentage for the United States

TABLE 1.1   Average Trade-to-Income Percentages, 1960-1995

Period	U.S. Total	U.S. Agriculture
	Percent	
1960-1969	9.4	47.1
1970-1979	14.8	64.3
1980-1989	18.7	85.6
1990-1995	22.1	86.4
1960-1995	15.6	69.1

Source: Derived from *Economic Report of the President*, Washington, DC, 1996.

increased from 9.4 percent during 1960-1969 to 22.17 percent during 1990-1995 (Table 1.1).[1] This suggests a significant opening of the economy. The comparable trade-to-income percentages for agriculture suggest an even more significant opening of the farm economy in the United States. The percentage for agriculture increased from 47.1 percent (during 1960-1969) to 86.4 percent (during 1990-1995).

Alternative measures of the openness of the U.S. financial sector to international transactions indicate that the U.S. financial sector has also became more highly integrated with international capital and money markets in recent decades. For example, the average annual percentage of U.S. net private asset outflows plus net foreign asset inflows (excluding official assets) increased from about 1.4 percent of U.S. gross domestic product (GDP) during 1960-1969, to 5.28 percent of GDP during 1980-1989, and fell slightly to 4.26 percent during 1990-1995. These are relatively small percentages, since they represent net asset flows.

A second measure of the increased level of international flows in U.S. financial markets is the percentage of financial assets held in financial and nonfinancial institutions by the "rest of the world" when compared with the overall flow of funds in the United States. That percentage was 3.4 percent in 1980, 5.5 percent in 1990 and 6.2 percent in 1994. Thus, there has been a rather steady increase in the proportion of foreign-held assets in recent years.

A related indicator of financial sector integration is growth in the volume of foreign transactions in stocks and bonds (public and private). The volumes and compound annual rates of growth of financial transactions during 1980-1995, for U.S. purchases of foreign securities and for foreign purchases of U.S. securities, are reported in Table 1.2. The growth rate for Treasury securities transactions was about 55 percent during 1980-1985, 30 percent during 1985-1990, and 10 percent during 1990-1995. Purchases of corporate securities increased at about half the rate of transactions in Treasury securities during 1980-1990, but was about twice the level of Treasury transactions during 1990-1995. Overall, foreign purchases of total U.S. securities grew at about a 45-percent rate during 1980-1985,

TABLE 1.2 Volume and Growth of International Securities Transactions

Year	U.S. Purchases and Sales of Foreign Securities	Foreign Purchases and Sales of		
		U.S. Treasury Securities[a]	Corporate Securities[b]	Total U.S. Securities
		$ billion		
1980	53.1	114	84	198
1985	n.a.	1,014	243	1,256
	(n.a.)	(55)[c]	(24)	(45)
1990	906.7	3,724	479	4,204
	(n.a.	(30)	(15)	(27)
1995	2,573.6	6,050	1,192	7,242
	(23)	(10)	(20)	(11)

[a] U.S. Treasury securities includes marketable bonds and notes.

[b] Corporate securities include stocks and directly placed bonds of corporations and issues of states and municipalities.

[c] Numbers in parentheses are 5-year, compound, nominal growth rates (in percent) for the 5 years preceding the year indicated.

Source: U.S. Bureau of Census, *Statistical Abstract of the United States: 1996.*

27 percent during 1985-1990, and 11 percent during 1990-1995. U.S. investor purchases of foreign securities grew at about 23 percent during 1990-1995. Foreign purchases of total U.S. securities grew at a compound rate of 27 percent, while U.S. purchases of foreign securities grew at 30 percent during 1980-1995.

These indicators illustrate the increasing trend of international financial flows—a trend that is expected to continue and possibly accelerate into the next century. This trend has been driven by both supply and demand factors. Issuers of securities (including governments) have encouraged the internationalization of their investor base in order to enhance the liquidity of their securities and to reduce their costs of funding. Investors have sought to stabilize their returns through portfolio diversification (Fukao 1993).[2] The conclusions one draws from the data on trade and financial sector transactions is that the U.S. economy has opened significantly in the post-1980 period, and that the trends will likely continue (if not accelerate) in the next century.

If concern exists over greater openness of the economy and financial markets, it is that increased openness leads to a higher degree of financial and economic volatility than would occur if the economy were more insulated. Is there evidence that such an effect on volatility exists in the United States? Shiller (1988) suggests that relatively little is known about the determinants of financial market volatility and that there is no proven theory of financial fluctuations. Yet, even without such evidence, it is generally agreed that financial market volatility can have far-reaching ramifications in terms of disruption of domestic economic activity and international asset flows.[3]

Gertler and Hubbard (1988) suggest that the impact of financial market fluctuations on business activity (the real economy) occurs primarily through two channels: fluctuations in the internal net worth of firms and fluctuations in the availability of bank credit. In the first channel, the implication is that financial volatility creates an unexpected change in the level of collateral, a redistribution of wealth, and a change in the level of investment (a real effect). Via the second channel, financial volatility creates an unexpected loss of access to bank loans as a source of financing, and the level of investment falls due to reduced liquidity. Economic shocks transmitted through these channels are expected to have implications for agricultural sector financing.

Work on international business cycles and volatility has generated evidence of economic connections between countries that serve to transmit aggregate fluctuations. These connections include the volatility of a country's balance of trade, the correlation of the trade balance with the country's level of output, the correlation of output and consumption across countries, and the volatility of prices of goods produced in more than one country. For example, Backus, Kehoe, and Kydland (1993) examine cross-country volatility, correlations and comovements of output, consumption, and other aggregates. They find that the cross-country correlations (comovements) of countries with the United States in output are generally larger than the corresponding correlations in consumption, investment, and productivity shocks (Table 1.3). By modelling the outputs of the countries as imperfect substitutes, they show that there is a terms-of-trade (price) effect where the relative prices of the traded goods may diverge. These data suggest that comovement is present, but that it varies by the country and aggregate indicator selected.

TABLE 1.3 International Comovements by Indicator and Country, 1970 to Mid-1990[a]

Country	Output	Consumption	Investment	Gov't. Purchase	Employment	Productivity
Australia	0.51	-0.19	0.16	0.23	-0.18	0.52
Canada	0.76	0.49	-0.01	-0.01	0.53	0.75
Japan	0.60	0.44	0.56	0.11	0.32	0.58
Europe	0.66	0.51	0.53	0.18	0.33	0.56

[a] Comovement is measured by the correlation of each country's aggregate variable with the corresponding variable in the U.S.

Source: Backus, D. K., P. J. Kehoe, and F. E. Kydland. 1993. "International Business Cycles: Theory and Evidence," Quarterly Review, p. 17

Table 1.4   Effects of Exchange Rate Regimes on Volatility and Comovement

| Country | Period | Exchange rate regime | Volatility: Standard Deviation | | Comovement: Correlation with same U.S. variable | |
			Terms of trade	Output	Output	Consumption
			Percent		Ratio	
Canada	1955-90	—	2.44	1.48	.71	.52
	1955-71	Fixed	1.19	1.38	.53	.59
	1972-90	Floating	3.05	1.54	.79	.48
Japan	1955-90	—	5.69	1.61	.20	.27
	1955-71	Fixed	2.29	1.93	-.07	-.02
	1972-90	Floating	7.12	1.19	.57	.36
U. K.	1955-90	—	2.64	1.48	.46	.35
	1955-71	Fixed	1.45	1.25	.15	.05
	1972-90	Floating	3.05	1.67	.57	.35
U. S.	1955-90	—	2.92	1.70	1.00	1.00
	1955-71	Fixed	1.26	1.23	1.00	1.00
	1972-90	Floating	3.79	1.94	1.00	1.00

Source: Adapted from Backus, D. K., P. J. Kehoe, and F. E. Kydland. 1993. "International Business Cycles: Theory and Evidence," Quarterly Review, p. 27.

In addition to the evidence of significant comovement in macroeconomic aggregates, there is also evidence that the shift from a fixed to a floating exchange rate regime led to an increase in both the level of volatility and comovement in and among these countries (see Table 1.4). The volatility of terms of trade and output both generally increased during 1972-1990. The correlations between output and consumption aggregates generally confirm also that the shift in exchange rate regime was followed by an era of increased comovement, suggesting greater international economic interdependence.

There is less evidence that financial markets have become significantly more interdependent. Studies of the correlation between equity markets in different countries provide mixed evidence on the coherence of international equity capital markets, although international stock markets appear to be more integrated than the corresponding economies (Bowden and Martin 1995). Yet, it is generally the view that international transmission mechanisms can play a major role in key situations—as evidenced by the U.S. stock market crash in 1987 and the growing importance of foreign transactions in stock exchanges.

TABLE 1.5 Key Changes in Global Capital Markets, 1940s and 1990s

The 1940s	The 1990s
The United States was the dominant economic super- power, and the formation of an international economic system could be determined by the United States and several of its close allies.	The United States is one of several economic superpowers, and reform of the international economic system now requires consideration of numerous views from many different nations.
Many leading governments practice exchange rate and capital controls to highly regulate financial markets.	Few leading governments practice exchange rate and capital controls, and deregulation has altered many financial markets.
A major portion of international capital flows was conducted between government (or quasi-government) agencies.	A major portion of international capital flows is conducted between private sector, profit-maximizing corporations.
International trade was "king," and international finance was the junior partner.	International finance is "king," and international trade often is the junior partner.
Computers and derivative instruments (e.g., interest rate swaps, futures, and options) had no influence on, and did not exist in, most global capital markets	Computers and derivative instruments have a major influence on, and exist in, most global capital markets.
It was a world of fixed exchange rates, and national regulators were able to cope with the international side of financial markets.	It is a world of "dirty" or managed floats, and national regulators are often unable to cope with the international side of financial markets.

Source: Derived from Belous.

Trends Toward the Globalization of Financial Markets

Belous (1982) identifies several basic changes that have altered the landscape in global financial markets since the 1940s (Table 1.5). The salient features of this changing environment include: adoption of more flexible exchange rate regimes, less-regulated financial markets, introduction of new derivative instruments, and an increasing flow of international capital between private corporations—a flow that follows the dictates of market forces.

Thus, powerful trends have been underway for decades. These trends have converted national financial systems into an integrated global system for the purposes of attracting savings, extending credit, making payments, and fulfilling the role of increasing international capital mobility. This broadening of financial markets has been driven by technological and entrepreneurial innovations. Improvements in communications technology (satellites, computers, and other automated systems) now tie together financial service businesses with trading centers and clients in a globally dispersed system. Domestic and international financial markets now have the technological capacity to perform the same functions: bringing lenders of funds in contact with borrowers, channeling the flow of scarce financial resources toward their most productive uses. The increased volume of currency, capital, and commodity flows are all facilitated through this rapidly emerging financial system.

According to Kaufman (1994), financial markets have undergone significant structural changes that make the financial system vulnerable to excess. The three primary forms of structural change that have contributed to this condition are: (1) the diminished role of traditional lending and investing institutions in determining the composition of investment and responses to market developments; (2) the trend toward "Americanization of finance," which encompasses a deregulation of markets and institutions, growth in securitization, greater use of new financial instruments (e.g., derivatives) and trading techniques, and growth in the presence of a group of portfolio managers who "roam the financial sphere" with a near-term focus; and (3) the development of an infrastructure that promotes credit creation outside the banking system and outside the purview of central banks. The growth of mutual funds, in particular, has contributed to the diminished role of traditional lending and investing institutions. Because these "risk pass-through institutions" have grown in importance, Kaufman expects the volatility of financial markets to increase in the future.

Kaufman (1992) also finds that globalization and deregulation of financial markets has provided both benefits and costs. The benefits are due to the increased mobility of capital seeking out the most profitable deployment of financial resources. The costs arise due to the increase in financial excesses and abuses - primarily due to the use of poor credit judgment and/or an inadequate understanding of the risks involved. Moreover, the safety and soundness of the international financial system increasingly depends on the development of a more forceful and internationally coordinated supervisory presence. Kaufman points to capital scarcity in the global economy as a condition that will continue to transform international financial markets into the next century.

Agriculture is undergoing a structural and technological transformation that makes it more sensitive to changing conditions in national and international financial markets. Both technological change and growth in farm business size have been increasing the capital requirements of agriculture and, thus, the demand for external financing. That expanding demand for investment capital implies that farms and agricultural businesses must increasingly compete for a larger share of their financial capital in the national and international marketplace. Not only will a larger volume and share of total capital (both debt and equity) come from these sources, but the terms under which that capital is supplied will be dictated by conditions in national and global financial markets.

In this context, the financial institutions serving U.S. agriculture are undergoing major changes and they are subject to the same set of economic and technological forces that have been reshaping national and international financial markets. We cite two examples of the trend toward greater linkages between the national financial market, the international financial market, and the financing of U.S. agriculture. Those examples are the funding of the Farm Credit System (FCS) and increased lending by foreign banks in the United States (i.e., transnational lending).

Farm Credit System Funding. The FCS is a government-sponsored enterprise (GSE) that has historically issued debt securities (notes and bonds) in the national capital and money markets to fund loans to U.S. agriculture (see Chapter 6). Due to "agency status" afforded to the FCS, it is able to issue those securities at a relatively small spread over the rate on comparable-maturity U.S. Treasury notes and bonds. It is generally accepted that this status has reduced the cost of funds to the FCS and has broadened the market of investors in FCS securities.

Although the process of issuing FCS securities has changed in recent years, the end-buyers of FCS securities has remained relatively unchanged over the past 10-15 years. Because the spreads over U.S. Treasury debt are quite small, GSE buyers generally include investors who are inclined (or required) to hold high-quality securities in their portfolios. This characteristic tends to produce a relatively stable investor base. The typical end-buyer distribution of FCS short- to medium-term (3-month to 5-year maturity), nonhighly structured securities (according to volume of issuance) has been dominated by state and local governments (about 30 percent), commercial banks (20 percent), and investment advisors and fund managers (30 percent). Foreign central banks and public bodies have historically accounted for only about 5 percent of total volume issued.

Demand for FCS securities reflects their acceptability in the national and the international financial marketplace. In recent years, foreign public bodies have become more active in buying GSE securities. In October 1996, the FCS issued its first global bonds. The $500-million issue of three-year callable bonds was reportedly oversubscribed by about 10 percent. Buyers of the bonds were about evenly distributed between buyers in the United States (36 percent), Asia (34 percent), and Europe (30 percent). It is believed that this successful inaugural issue of FCS global bonds has demonstrated that, in addition to a strong national market, there is a viable global market for these securities. Further FCS funding through international capital markets is envisioned.

Transnational Lending. The emergence of increased lending by foreign banks and their subsidiaries in the United States is symptomatic of a changing competitive banking environment. U.S.-owned banks increasingly operate in foreign countries and foreign-owned banks in the United States. Increased foreign lending also suggests a wider array of financing alternatives for nonfinancial sector

TABLE 1.6 Number, Assets, and Market Share of Foreign Banks in the United States, 1980-1993

Year	Banks	Assets	Share U.S. Market
	Number	$ billion	Percent
1980	462	262	13.0
1983	598	388	15.1
1985	658	507	16.7
1987	697	665	19.4
1989	732	814	21.2
1991	754	970	23.7
1993:II	700	948	22.4

Source: Nolle, p. 4.

clients (including agriculture). A recent assessment of the market share of foreign-owned banks (subsidiaries, branches, and agencies) in the United States suggests that they have grown dramatically in the 1980s and early 1990s (Nolle 1984). The 1980-1993 era saw an increasing number of foreign banks in the United States, an increasing level of assets in those banks, and increasing market share (Table 1.6)

Corresponding with the increase in foreign bank growth has been a rapid increase in commercial and industrial (C&I) loans by those banks to U.S.-based businesses, some of which are agribusinesses. Nolle estimates that the foreign share of the C&I loan market increased from about 19 percent in 1980 to about 47 percent in 1993. However, the growth of foreign bank involvement in the United States loan market appears to have reached a plateau in the early 1990s. Various reasons have been cited for the growth in foreign bank lending: the entry of foreign banks into the U.S. market to facilitate the trade and direct investment needs of their home country clients (Budzeika 2992), a cost-of-capital advantage enjoyed over U.S. banks until 1990 (Zimmer and McCauley 1991), and temporary cross-country differences in bank capital requirements as a result of the Basle Accord in 1988 (Frankel and Morgan 1992; Wagster, Cooper, and Kolari 1994). The study by Nolle indicates that foreign banks have "paid a price in terms of efficiency and profitability." Although there is evidence of an increasing integration of the "global financial community," banks still have significant problems competing in these new foreign markets.

The increase of transnational lending in U.S. agriculture can be illustrated more specifically with the example of Rabobank, N.A. The Rabobank Group is a cooperative bank network headquartered in Utrecht, the Netherlands. Rabobank launched its North American operations in 1980, and currently operates offices in six U.S. cities. The client base includes input sectors (farm equipment, agricultural chemicals, and fertilizers), agricultural producers, processors, distributors, marketers (wholesale and retail), importers and exporters, and commodity trading companies.

By 1996, the Rabobank N.A. portfolio had grown to about $13.7 billion in funded and contingent assets and total credit commitments amounted to $16.1 billion—54 percent to agribusiness and 46 percent to nonagribusiness (Ziengs 1996). The agribusiness credit volume is broadly diversified among commodity groups: grains, livestock and poultry, dairy, horticultural crops, sugar, coffee, cocoa and cotton, food service and distribution, agricultural equipment and chemicals, seeds, and agricultural export financing.

The initial incentive for Rabobank to enter the U.S. agribusiness market in 1980 was the perceived opportunity to focus on the agribusiness market - a sector where Rabobank has historically had considerable knowledge. This provided Rabobank with a competitive advantage over other foreign banks also establishing branches in the United States at the time. To sharpen this advantage, Rabobank hired experienced U.S. agribusiness bankers who knew the industry and the players. This competency, combined with the fact that U.S. banks had begun to reduce their emphasis on agriculture and agribusiness, provided Rabobank with several timely business opportunities in the sector. From the beginning, Rabobank has also enjoyed a competitive advantage over other banks due to its Triple-A credit rating, which provides it with access to lower-cost funds.

Changing Linkages Between Exchange Rates, Interest Rates, and Agriculture

Macroeconomic policies can influence agriculture directly and indirectly through their effects on interest rates and exchange rates. Changes in exchange rates may influence agriculture indirectly by affecting the export and import markets for agricultural commodities. Interest rate fluctuations directly affect the terms of agricultural loans and indirectly influence exchange rates. Thus, in relatively short periods of time, changes in macroeconomic policies can shift interest rates and exchange rates, and cause the terms of agricultural loans to change. Thus, various relationships between macroeconomic policies and conditions in financial markets are key to understanding how macroeconomics can influence agriculture, now and in the future.

Interest rates and exchange rates are so closely related that macroeconomic policies cannot affect one without affecting the other. Domestic and foreign interest rates and exchange rates are determined by market forces that enforce the principle of *interest rate parity*. Interest rate parity requires that the rate of return on comparable securities (net of expected foreign exchange gains and losses) be equal in all countries. That is, the net rate of return to the investor from any foreign investment is equal to the interest earned plus (or minus) the forward premium (or discount) on the price of the foreign currency involved in the transaction. For example, in a country where interest rates are higher than in the United States, the foreign currency will sell at a discount in the forward exchange market.

To illustrate this point, one-year Japanese government bonds yielded less than 1 percent while one-year U.S. Treasury bills yielded close to 6 percent in 1996. Why didn't investors simply choose to sell their Japanese securities and buy U.S. Treasury bills? The explanation is found in the forward exchange market, where the expectation is that the yen will appreciate against the dollar and compensate investors for the lower interest rate on Japanese securities. Ignoring tax considerations and bid-ask spreads, the equation for interest rate parity between the Japanese and U.S. securities can be written as,

$$(1 + i_J) = (1 + i_{US}) \times (1/Spot) \times (Future)$$

where i_{US} is the interest rate on the one-year U.S. Treasury bill, i_J is the interest rate on the one-year Japanese Treasury bill, Spot is the current value of one dollar in terms of yen, and Future is the value of one dollar delivered twelve months from now in the forward exchange market. Interest rate parity indicates that if the current exchange rate is 105 yen/dollar, the futures market must place a future value on the yen of approximately 100.05 yen/dollar. If the futures market placed a significantly higher or lower value on the yen, an arbitrage opportunity would exist. Due to the high level of efficiency of international currency markets, these conditions could not persist, and arbitrage would quickly bring current and future exchange rates in line with the observed interest rate differential. Thus, due to the potential for arbitrage, the difference between current and future exchange rates will always reflect interest rate differentials, and interest rates and exchange rates will move together due to market efficiency (Grabbe 1986).

Several factors in addition to interest rate shifts can affect current and future exchange rates, including balance-of-payments positions of countries, speculation, domestic economic conditions, and the intervention of central banks (Rose 1994). When a country generates a trade deficit, it becomes a net demander of foreign currencies and it is forced to sell its own currency to pay for imported goods and services. Therefore, a persistent trade deficit can lead to the depreciation of a nation's currency relative to other currencies. Domestic monetary policy, fiscal policy, and political conditions may have favorable or adverse effects on exchange rates. Inflation is a particularly significant factor, since an unexpected increase in the domestic rate of inflation will cause the domestic currency to fall relative to other currencies. Conversely, a country that successfully reduces its inflation rate will cause the domestic real interest rate to rise, and with it the exchange value of the domestic currency. Finally, central banks may intervene in currency markets to stabilize their currency. Such interventions can have currency market and money market effects if currency transactions are not "sterilized" by appropriate, offsetting monetary policies that absorb changes in reserves.

Macroeconomic Policies and Productivity Shocks

A host of macroeconomic forces affects interest rates, exchange rates, and agriculture. Three of the forces given special attention in the economics literature are monetary policy, fiscal (and farm) policy, and productivity shocks. These three forces have been identified as major contributors to economic cycles and long-term shifts in interest rates and exchange rates.

Monetary Policy and Inflation. Monetary policy has been hypothesized to affect the real economy through several alternative channels: an interest rate channel, an exchange rate channel, and a credit channel (Mishkin 1995). The effect of monetary policy via the interest rate and exchange rate channels is illustrated by, for example, the Federal Reserve Bank selling U.S. Treasury securities to reduce the money supply. The purchase of those securities by the public removes deposits or free reserves from the banking system and causes real interest rates to rise. The increase in real interest rates causes an influx of foreign capital and the value of the U.S. dollar begins to rise. Because of currency appreciation, net exports decline until domestic price levels adjust to the reduced level of money supply. The net result is that a decrease in the supply of money can have real, short-run effects on agricultural and other exports through the exchange rate mechanism. The credit channel has its effects on financial markets as a result of "agency problems" in credit markets. Because of asymmetric information about borrowers, a reduction in money supply reduces bank deposits, which in turn reduces bank loans, investment, and output.[4]

With regard to the interest rate channel, it has been recognized for some time that the yield curve (which graphically illustrates the term structure of interest rates at a point in time) contains information that can be useful as an indicator of future economic developments. This occurs because the yield curve reflects monetary policy instrument settings, as in the level of short-term interest rates and the market's general view on future short-term rates. Therefore, it also reflects future rates of growth and inflation. For example, an easing of monetary policy will tend to lower short-term interest rates and cause the yield curve to steepen, leading to an increase in the prospective rate of growth and inflation. Similarly, a shift in market expectations toward higher rates of inflation or growth will tend to raise long-term rates and lead to a steeper yield curve.[5]

The shortrun effectiveness of monetary policy (i.e., the ability to generate real sector effects) rests on the assumption that domestic prices and trade are "sticky" (Dornbush 1981). Capital flows are thought to adjust rapidly to the new interest rates, which in turn causes a rapid shift in the exchange rate. If some prices are sticky, the inflation rate will not adjust rapidly enough to offset the shift in exchange rates. Thus, the key assumption behind the anticipated effectiveness of monetary policy is that international capital responds faster than domestic prices. The general implication is that a shift in monetary policy can have "real effects" on the terms of trade and trade flows. Stamoulis and Rausser (1988) point out that agricultural prices shift faster than other traded goods and, as a consequence, the price of agricultural goods can be particularly volatile due to "overshooting."

The influence of commodity price changes on the general rate of inflation has received the attention of economists because of the perception that changes in those prices have a direct bearing on inflation (and ultimately on interest rates). Commodity prices have been linked to inflation through two channels: they respond to general economic shocks more quickly than other prices, and some changes in commodity prices reflect idiosyncratic shocks (e.g., weather-related changes in the supply of agricultural products) that are passed on in the form of higher overall prices (Furlong and Ingenito 1996). In this way, commodity prices have been historically viewed as leading indicators of future inflation. The argument for quickly responding commodity prices is that the prices of commodities are determined in highly competitive auction markets, and they tend to be more flexible. The idiosyncratic shock argument rests largely on the view that a direct shock to the supply of an agricultural commodity will result in higher food prices, which will affect prices overall (if the prices of other consumer goods are sticky).

Furlong and Ingenito indicate that the link between commodity prices and inflation has changed dramatically. During the 1970s and 1980s, commodity prices tended to be relatively robust, leading indicators of inflation, but in the latter 1980s and 1990s they have been relatively poor "stand-alone" indicators. The change in the role of commodity prices is attributed to movements in commodity prices that increasingly reflect idiosyncratic supply shocks.

Fiscal and Farm Policy. In the 1980s, tight monetary policy (slow growth in money supply and higher interest rates) was accompanied by loose fiscal policy (growing federal expenditures as a percent of GDP). Teigen and Shane (1995) show that federal government spending increased in the 1980s to the point where the U.S. deficit was about 3 percent of GDP in 1993, and exceeded comparable levels in the United Kingdom, France, and Japan.[6] As a result of this escalation, the budget deficit became the focus of several empirical studies of fiscal policy, exchange rates, and the trade deficit.

A key factor underlying the effectiveness of fiscal policy is the degree to which individuals compensate for governmental deficits (public spending) with additional private saving. If budget deficits do not induce an offsetting increase in private saving, interest rates are expected to rise with government deficits.[7] Historical evidence indicates that consumers do not sufficiently increase their savings to fully compensate for increases in the budget deficit. As a consequence, the issuance of additional governmental debt drives up interest rates and attracts an influx of foreign capital. After the value of the domestic currency increases, exports fall. In the 1980s, the budget deficit and the trade deficit were termed the "twin deficits" due to the conventional view that loose fiscal policy was driving up the trade deficit.

There remains an ongoing debate about the relationship between budget deficits (rather, changes in budget deficits) and changes in exchange rates (Feldstein 1995). This is because there are both direct and indirect effects of deficit reductions, which act in opposing directions on exchange rates.[8] Deficit reduction can lead to a weaker exchange rate because the deficit reduction causes a *direct effect* on interest rates and exchange rates. This occurs since the deficit reduction reduces the demand for loanable funds. If the deficit is reduced, the demand for funds declines and both interest rates and exchange rates tend to fall. The mechanism by which exchange rates adjust is through the actions of private investors who sell low-return domestic assets, use the domestic currency to buy foreign currency, and use the foreign currency to buy higher-return foreign securities. Thus, a decline in domestic interest rates reduces the demand for the domestic currency in the market for foreign exchange and causes the exchange rate to depreciate.

Alternatively, deficit reduction can lead to a stronger exchange rate when deficit reduction leads to an increase in the demand for funds by private investors. This occurs through three *indirect effects*, which induce private investors to increase their demand for domestic securities relative to foreign securities. Those three indirect effects include a lower rate of expected inflation, a lower foreign exchange premium, and an increase in the expected rate of return on domestic securities (Melvin 1992).

Whether exchange rates rise or fall depends on the relative size of the *direct* and *indirect effects*. Hakkio suggests that the *indirect effects* (which increase the exchange rate) tend to dominate the *direct effects* when deficit reduction is "credible, long-term and sustainable" (p. 27). That is because only under those conditions will the effect of deficit reduction on inflation expectations, the risk premium, and expected rate of return be significant and induce private investors to hold relatively more domestic securities. New empirical results indicate that the mode of deficit reduction (a reduction in spending versus an increase in taxes) and the country's policy with respect to monetizing the debt are important determinants of the effect (Hakkio). Hakkio finds a small effect of deficit reduction (direct effect) on the exchange rate in the G-7 (the seven major industrialized) countries and similar results in other OECD countries. The analysis suggests also that the exchange rate strengthens more (weakens less) when deficit reduction is accomplished by reducing spending rather than by increasing taxes. These direct effects of deficit reduction suggest mixed results in that only seven of the eighteen countries analyzed showed significant response of exchange rates to changes in the deficit. Hakkio's analysis indicates that the indirect effects are generally significant, but small.

Evans (1986) did not find a causal relationship, while Feldstein (1986) and Rosensweig and Tallman (1993) found a causal link from budget deficits to exchange rates and the U.S. trade deficit. In a sector-specific study of the influence of fiscal policy on agriculture, Devadoss and Chaudhary (1994) concluded that even anticipated shifts in budget deficits can affect the level of real agricultural output. The overall evidence suggests that fiscal policy is nonneutral in its effects. However, the effectiveness of fiscal policy and the ability of the government to fine-tune the economy with fiscal policy both remain open questions.

Recent changes in farm policy, as part of the fiscal policy framework of agriculture, also represent a significant development that is expected to influence the financing of agriculture into the next century. Passage of the Federal Agricultural Improvement and Reform Act of 1996 (FAIR) makes major changes in the substance and implementation of agricultural policy in the United States. FAIR provides for a suspension of price supports for food grains, feed grains, and cotton, and complete farmer planting flexibility among a wide variety of crops, including the option not to plant and yet receive a payment. FAIR also removes the Secretary of Agriculture's ability to require producers to idle land in order to receive farm program payments. Although this is a major departure from past farm programs, there are indications that the degree to which U.S. producers will adjust their production decisions in the near term may be somewhat limited, partly due to the phasing in of the new program and the need for time until producers recognize the price signals implied by the new market environment (Food and Agricultural Policy Research Institute 1996).[9] Thus, a dampened initial supply response is anticipated. However, FAIR will allow producers to respond more quickly than in the past to market price changes. It will also allow for changes in a broader range of market prices than occurred previously. The change in rules for participating in the farm program will allow the sector as a whole to respond more quickly to relative market price adjustments.

Productivity Shocks. Productivity shocks are forces that shift the expected returns of labor and/or capital. Thus, a capital gains tax cut that increases returns to capital could be considered a productivity shock to capital. Similarly, a breakthrough in biotechnology (e.g., genetically engineered seeds and animals) could be considered a productivity shock to the returns from land and labor in agriculture. Generally, if factor productivity increases, financial capital is expected to flow to the countries whose productivity is increasing most rapidly. The flow of capital across international borders will influence interest rates, exchange rates, and trade flows.

Productivity shock models are often called "real business cycle" (RBC) models, since a combination of beneficial shocks (e.g., technological innovations) and adverse shocks (e.g., an oil embargo) are used to explain observed economic cycles. RBC models can help explain a significant portion of economic volatility with external productivity shocks, as opposed to relying on price stickiness and shifting government policies (Kydland and Prescott 1982). The RBC approach points out how exchange rate and commodity price volatility could jointly stem from shifting factor productivity, as opposed to fiscal and/or monetary policy.

Backus, Kehoe, and Kydland (1994) explain a portion of exchange rate shifts and trade deficits based on productivity shocks. According to their analysis, a positive productivity shock can cause increased consumption and a flow of capital into the affected country, resulting in a temporary decrease in net exports. These results are an attempt to explain the fact that fluctuations in national real growth rates tend to be negatively correlated with national net exports. While their model does not explain the full extent of exchange rate volatility, it does demonstrate how interest rate and exchange rate volatility can affect agriculture even when fiscal and monetary policy remain stable.

We conclude that monetary and fiscal policies and productivity shocks provide alternative explanations for market volatility. If prices are sticky, monetary policy can induce farm income volatility. If savings levels do not respond to government deficits, fiscal policy will have real effects on farmers and agricultural lenders. Even when governments pursue consistent policies, real productivity shocks can cause changes in interest rates and in financial markets. The effect of these developments on individual farmers and farm lenders will depend on their degree of exposure to changing interest rates and on how shifts in exchange rates influence commodity prices.

Exchange Rates and Agricultural Exports

The effects of exchange rate movements on agriculture can be evaluated by studying the volatility of exchange rates in the 1970s and 1980s (Chambers 1988). During that period, shifts of exchange rates were followed by opposing shifts in agricultural exports. Figure 1.1 illustrates the relationship between a trade-weighted index of real exchange rates and U.S. agricultural exports (Stallings 1992; USDA various).

After devaluing the dollar in 1971 and 1973, the United States allowed the dollar to float in international currency markets. Market forces caused the dollar to move downward throughout the 1970s and agricultural exports exhibited spectacular growth. The process was reversed in the 1980s when the dollar's rapid rise was followed by a collapse of agricultural exports. In the latter 1980s, U.S. agricultural exports rebounded as the dollar began a long-term, secular decline. Although the dollar has depreciated slightly in the 1990s, the recent surge in agricultural exports is usually not attributed to exchange rate movements (Coyle 1996).

The strength of the dollar in the mid- and latter 1990s has created problems for other countries. An illustration of this is unfolding during mid-1997 in Southeast Asia. Several Southeast Asian countries have historically pegged their currencies to the U.S. dollar. They include Thailand, Singapore, Indonesia, Taiwan, Malaysia, and the Philippines. However, with the collapse of Thailand's finan-

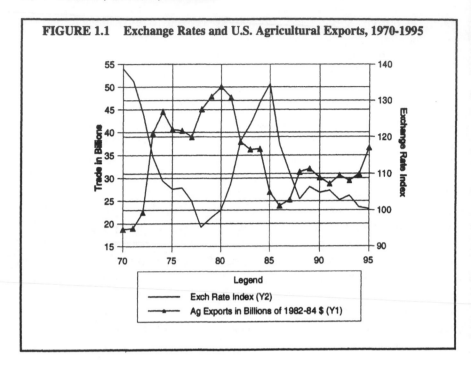

FIGURE 1.1   Exchange Rates and U.S. Agricultural Exports, 1970-1995

cial sector and related economic problems, and the subsequent erosion of the baht by 20-25 percent against the dollar, the currencies of other Southeast Asian countries have also come under attack (Wall Street Journal 1997). For example, the Indonesian rupiah has been pegged to a 12-percent trading band around the dollar in the past, but widespread selling has forced it lower. The Southeast Asian governments face a dilemma: to reduce interest rates in order to stimulate economic growth, or to continue to peg their currencies to the dollar, which requires that they keep interest rates (and demand for their currencies) high. In this currency crisis, traders have voiced concerns that the inability of the central banks in the region to visibly intervene and stem the fall of their currencies is signalling an end to the era of the inflexible exchange rate and that a period of increased currency volatility is emerging.

What does this volatility in currency markets signal for North American agriculture? First, it means that floating currencies in Southeast Asia and elsewhere will lead to a downward adjustment in the exchange rates of trading partners. One consequence of this rate adjustment will be a temporary economic slowdown in the region. Since an economic slowdown affects trade, there may be less demand for U.S. exports (including agricultural commodities) in that region during the next few years. Second, the depreciation of Asian currencies relative to the dollar means that U.S. exports will become less competitive in that region and demand for U.S. exports will also be driven downward. Thus, due to the

Southeast Asian currency crisis, there will be an adverse impact on U.S. trade in the region. This may hurt U.S. agricultural exports, both in the region and internationally, as the currencies realign. If the "signal" of these events is for greater currency volatility in the future, there may be even greater long-term volatility in store for international trade in agricultural commodities.

Empirical Studies. Schuh's (1974) explanation of how an overvalued currency can act as a tax on agricultural exports and a subsidy on imports prompted several empirical studies that attempted to quantify the effect of exchange rate movements on exports and domestic commodity prices. Estimates of export elasticities have reflected sensitivity to exchange rates and the level of government intervention (Roe, Shane, and Vo 1986). Those estimates suggest that a 1-percent shift in exchange rates would lead to a change in exports of about 1 percent (Bredahl, Meyers, and Collins 1979). Chambers and Just (1981) also examined data from the 1970s and concluded that a 10-percent exchange rate adjustment would cause an 8-percent shift in the domestic price of wheat. Chambers and Just (1979, 1981) also argued that, due to cross-price and income effects, the elasticity of commodity price with respect to changes in the exchange rate could be greater than 1.0 for corn and soybeans. These estimates of price elasticity are not precise, but they do provide an indication of how sensitive domestic grain prices were to exchange rates by the end of the 1970s.

In the early 1980s, the value of the dollar rose substantially as tight U.S. monetary policy was coupled with loose fiscal policy. By this time, the level of asset turnover in U.S. financial markets exceeded the dollar value of international trade (Mussa et al. 1994). Thus, the volume and speed of capital flows allowed for rapid movements in exchange rates. High real interest rates coupled with an increased budget deficit caused a large flow of investments into dollar-denominated securities and a dramatic increase in the dollar's value. The dollar's rise and a global recession in the early 1980s combined forces to dry up the demand for U.S. agricultural exports (Battan and Belongia 1984).

During the mid-to-late 1980s Shane (1990) found that a 1-percent shift in the agricultural trade-weighted exchange rate caused exports to change by 1.35 percent, and that approximately 25 percent of the change in exports could be explained by adjustments in the exchange rate. Based on the empirical evidence, we can conclude that exchange rate shifts have induced commodity price shifts of about the same magnitude.

Future grain prices may be increasingly sensitive to exchange rate movements due to the elimination of price support programs as stipulated in FAIR.[10] With the passage of the FAIR Act, the impact of global price changes will not be cushioned by the government programs of the past. In addition, the international grain trade is becoming more like intrafirm trade. The internationalization of commodity trading firms, coupled with improved transportation and a reduction in trade barriers, will contribute to more efficient commodity trading. The implication is for faster changes in commodity prices in response to changes in interest rates and exchange rates. While macroeconomic forces are expected to continue to cause interest rate and exchange rate volatility, the globalization of agricultural markets and financial markets will cause financial market volatility to be translated into income volatility for North American agriculture more quickly than in the past.

Financial Markets and Risk

Financial institutions serving agriculture operate within a system of financial markets characterized by various elements of risk. For example, financial intermediaries are exposed to several major types of investment risk. *Interest rate risk* (or market risk) is generally defined as the potential variation in returns caused by unexpected changes in interest rates. As such, the variation in returns may be increased by characteristics of the underlying asset or liability (e.g., the stated interest rate or maturity). Thus, changes in market interest rates imply changes in the market value of securities and loans and, therefore, variations in the rate of return on those assets. Second, financial intermediaries are exposed to *reinvestment risk*, which is the risk that cash flows from a financial asset (e.g., a mortgage) will be reinvested in lower-yielding assets in the future. Thus, the actual yield falls short of the expected yield due to unexpected changes in market interest rates. Financial institutions and other investors also face *default (or credit) risk* due to the probability that a borrower will fail to meet some or all of the promised payments of principal and/or interest. Financial institutions and investors in international securities and loans also face *currency (or exchange rate) risk* due to the possibility of an unfavorable change in the value of the foreign currency in which their investments are denominated.

Due to the broad set of risks involved, financial intermediaries (particularly depositories) depend on public confidence. That confidence can be maintained at two levels: (1) at the *aggregate level* by developing and enforcing regulatory institutions (e.g., capital adequacy standards) and providing insurance institutions that ensure the safety and soundness of the financial system, and (2) at the individual *institution level* through management efforts to assess the level of risk exposure and to control the level of exposure within acceptable levels.

Risk-Based Capital Standards. Bank regulators from ten major banking nations met in Basel, Switzerland, in 1988 to agree on international standards for capital adequacy with respect to foreign exchange risk and commodity trading risk. The purpose of these regulations was to require that banks have adequate reserves to cover their anticipated losses. The international agreement was that all financial institutions should meet a minimum capital standard of 8 percent of their risk-weighted assets.

Agricultural banks meet this standard quite easily given that the average capital position of small banks generally exceeds that level. However, no regulations have been developed specifically to monitor the risks of agricultural lenders, which derive indirectly from exchange rate fluctuations or more directly from commodity price volatility.[11]

Insurance Systems. Governments have increased their efforts in recent years to ensure the safety and soundness of banks and other financial institutions to protect the public's funds. The Federal Deposit Insurance Corporation (FDIC) was developed in the 1930s to provide insurance coverage of small deposits. In recent years, there has been a growing recognition that government insurance must try to avoid distorting risk-taking decisions of managers of private financial institutions. Federal deposit insurance has protected small depositors, but it has also led many banks and thrift institutions to take greater risks—knowing that the insurance premiums paid by all institutions was the same, regardless of the risk exposure. The result was that the safer depository institutions subsidized the riskier ones.[12] In recognition of this problem, the FDIC has moved to institute a risk-rating system according to which each institution pays a deposit insurance premium based on the risk category into which it falls.

The Farm Credit System Insurance Corporation (FCSIC) became fully operational on June 1, 1993. FCSIC was established to ensure the prompt payment of most of the System's liabilities following the near collapse of the System in the mid-1980s. The reserves of the FCSIC Insurance Fund are funded through a $260-million transfer from a preexisting federally financed fund and the premiums paid by System banks and through income generated by its own investments. The FCSIC premiums are currently based partly on credit risk, but not on the interest rate risk or other risk exposures of the banks.[13] The FCSIC is authorized to assist troubled System banks or the direct lender associations (the ACAs, FLCAs, etc.). Through 1996, the Insurance Fund had not experienced a significant loss through its insurance operations due to generally improving financial conditions at the System banks (U.S. Government Accounting Office).

However, as the Agricultural Credit Act of 1987 established the FCSIC, it also restructured the Farm Credit System so that the direct lender associations became more directly responsible for lending decisions and managing credit risk within their portfolios. Thus, the question before the FCSIC is how to assess the capital and risk positions of the associations and extend the insurance system to cover those entities into the future. The implication is that the FCSIC will also need to consider the implementation of a broadened risk-rating and risk-based insurance premium system for the associations. The challenge is that this occurs at a time when the System is continuing to go through a period of significant merger and consolidation activity.

Innovations in Risk Management. Entrepreneurial (financial) innovations such as interest rate swaps, financial futures, and options contracts have added to the array of tools that can be used to reduce the costs of financing and to manage interest rate risk and currency risk. Although the use of these specific instruments and strategies is largely confined to large financial intermediaries, the popularity of their use is apparent from the rapid growth in national and international volume of these transactions.

Emerging Challenges for Agricultural Lenders

In the context of an increasingly complex set of macroeconomic factors and international linkages, agricultural lenders are challenged to effectively manage their institutions to maximize shareholder value, while meeting (or exceeding) the expectations of an increasingly sophisticated and more diverse set of clients. This strategy entails managing the interest rate risk, credit risk, and covariant risks that are endemic to lending institutions. It also entails managing more generic risks faced by all businesses - risks arising from a rapidly changing technological, regulatory, and competitive environment.

Interest Rate Risk

Among the most visible aspects of the macroeconomic instability of the past twenty years has been the behavior of inflation and interest rates. Consumer Price Index (CPI) inflation escalated sharply in the late 1970s, then declined to a relatively low and stable level in the early 1980s. Interest rates followed inflation, rising through the late 1970s, then declining secularly through the 1980s, as expectations were for initially rising, then falling inflation.

Through this period, volatile interest rates affected the earnings, market value, and liquidity of financial and nonfinancial institutions. The "savings and loan crisis" of the 1980s was partially a result of S&Ls funding their fixed-rate mortgages with deposits that ultimately proved to have variable rates. Market interest rates rose sharply following the Federal Reserve Board decision in October 1979 to reduce the growth rate of the money supply and decontrol interest rates. Deposit rates were gradually deregulated and the S&Ls experienced widespread disintermediation and reduced (or negative) net interest margins. The Farm Credit Systems (FCS) near-death experience in the mid-1980s was also closely related to the changing rate environment. Solvency of the FCS was threatened not only by adverse portfolio credit developments (associated with declining agricultural exports, farm incomes, and farm real estate values), but also from mismatched assets and liabilities and inadequate pricing policies. The FCS issued substantial amounts of high-cost, fixed-rate, noncallable debt during 1979-1981 to fund loans that were written with variable rates, or had fixed rates with no prepayment penalties. Because those loans were priced using variable (average) costs, rather than the marginal cost of funds to the System, farmers in the late 1970s found themselves in a position to buy FCS bonds with yields significantly greater than the rates they were paying on their farm loans. As market interest rates declined through the 1980s, FCS loan yields declined and net interest margins fell sharply. These events led to a recognition of the need for improved measurement and control of interest rate risk in the System through asset/liability management.

Interest rates played a major role in the agricultural credit crisis of the mid-1980s. Leveraged producers earnings were squeezed when interest rates rose in the early 1980s. High real interest rates contributed to a stronger dollar, reduced exports, and the devaluation of many classes of assets (including farm real estate). While interest rate risk is sometimes viewed by financial institutions as separate and distinct from credit risk, they are, in fact, related and must be addressed through a comprehensive risk management framework.[14]

Managing Interest Rate Risk. Interest rate risk arises from repricing mismatches, pipeline risk, prepayment risk, interest rate caps and floors, volatile spreads, and liquidity risk. Repricing mismatch refers to differences between the repricing characteristics of assets and liabilities (e.g., funding of variable-rate loans with long-term, fixed-rate liabilities). "Pipeline" refers to loans at various stages of the origination process. Interest rate caps and floors are limits placed on rate movements and are embedded in loan products such as adjustable-rate mortgages (ARMs). Spread risk refers to changes in the relationship between a lenders cost of funds and the lending rate.

Unanticipated spread changes may also be caused by macroeconomic phenomona and credit cycles. For example, commercial banks and the FCS experienced spread "compression" during 1988-1990, when interest rates increased and the yield curve flattened. Spreads between short- and long-term securities subsequently widened to unprecedented levels when interest rates fell and the yield curve steepened during 1992-1993. This allowed commercial banks and the FCS to generate record earnings and recapitalize themselves. By 1996-1997, credit spreads had narrowed to relatively low levels throughout the economy. Liquidity risk refers to the inability to neutralize a position by liquidating assets. Development of the interest rate swap market has greatly improved the ability of financial institutions to alter interest rate risk positions quickly and in large volumes with low transaction costs.

For the purpose of illustration, the process of interest rate risk management used at AgriBank, FCB is shown in Figure 1.2. An asset/liability simulation model is used to estimate the banks market value of equity (MVE) and net interest income (NII) under alternative interest rate scenarios, providing profiles of MVE and NII risks. The simulation model also provides "repricing gap" profiles under alternative assumptions regarding prepayment behavior, and a current "operating plan" forecast of bank earnings. Inputs to the simulation model include the banks "current position" (balances, yields, and run- off schedules for various classes of assets, liabilities, and off-balance sheet transactions), target balances, a funding strategy, interest rate scenarios, and prepayment functions. Simulations are performed monthly. *Ex post* validation of market-value risk estimates is performed following major interest rate movements.

FIGURE 1.2 Measurement of Interest Rate Risk

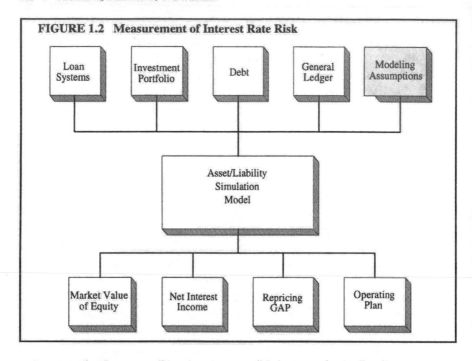

As strategies for controlling interest rate risk have evolved, they have increasingly included the introduction of new loan products (in some cases incorporating the use of derivative instruments). For this reason, the array and use of variable-rate loan products has increased significantly since 1980. For example, the share of farm loans with floating rates through commercial banks increased steadily from about 15 percent in 1980 to about 75 percent in 1995 (Walraven and Carson 1996).[15] This significant increase is consistent with the greater volatility of the cost of bank funds in a deregulated interest rate environment. Various loan products have been introduced in recent years by commercial banks and the FCS, including adjustable-rate loans and mortgages with rate caps and floors, loans with rates indexed to the prime rate, the Treasury bill rate or LIBOR, loans with prepayment penalties, etc. These new loan products have been adopted by the major financial institutions serving agriculture in order to protect their spreads from unexpected increases in market rates.

The introduction of some of these loan pricing options necessitates the continuous reevaluation of the costs of options embedded in loan products. These costs change not only because the yield curve shifts or market supply and demand for financial options changes, but also because the behavior of farmer-borrowers changes. As information about interest rates and option values has become more readily available and better understood by farmers, they have become more aware of the choices available to them for exercising embedded loan options (e.g., prepayment and product conversion options). As a reflection of this increasing borrower sophistication, the speeds of agricultural mortgage prepayment reached historically high levels as interest rates dropped during 1992-1993, causing the hedging costs for freely prepayable or convertible fixed-rate loans to rise. This led to changes in product features and the introduction of new products (e.g., varying prepayment and conversion options in fixed-rate loan products). Farmer-borrowers can now choose the level of prepayment optionality they desire.

The use of derivatives in commercial banking has increased in response to past interest rate volatility. The primary derivative instruments include interest rate futures, options on futures, and swaps. Banks can use derivatives in generally the same ways as investors: to hedge risk, to speculate on anticipated market movements, to adjust the financial characteristics of portfolios quickly and cheaply, and to arbitrage price discrepancies in financial mrkets (Becketti 1993). Commercial bank use of the derivatives market has been driven largely by money center banks, which serve as intermediaries for swaps and forward contracts in the over-the-counter market in order to generate fee income. The incentive for banks as "end-users" of derivatives flows from their traditional financing activities, which leave them exposed to financial market risks. For example, when a bank creates a fixed-rate mortgage, the exposure to interest rate risk is also increased. The bank could enter into an interest rate swap (to convert its fixed-rate asset into a variable-rate asset that matches its variable-rate funding sources) or it could hedge some of its interest rate risk in the futures market.

Covariant Risks

Due to the increased complexity and scope of risk considerations, agricultural lenders will need to apply portfolio risk concepts to better assess whole-firm measures of risks, capturing the important correlations between interest rates, spreads, and agricultural credit risks (i.e., covariant risks). Since shocks to interest rates, exchange rates, commodity prices, and inflation rates all act together to produce risks for farmers and lenders, the significance of these risks will depend on how they covary with each other. The correlation between agricultural prices and interest rates is particularly important, since this relationship will influence how successful banks will be in managing their risk exposures.

If *nominal* interest rates rise due to domestic inflation rates (as in the 1970s), farmers may be in a position to absorb interest rate risk. Broad-based inflation causes commodity prices and land values to rise and enhances the nominal cash flow of farmers and their net worth. As a result, bank loans become more secure in terms of debt service coverage and collateral coverage and interest rates on agricultural loans potentially could be increased without creating a significant increase in loan defaults.

If *real* interest rates rise due to restrictive monetary policy, the ability of farmers to absorb the higher debt service payments could be substantially reduced.

The higher real interest rates attract an influx of capital into the United States and put pressure on the dollar to appreciate. While the exchange rate is affected by a host of influences, there is evidence of a fairly strong historical correlation between real interest rates and real exchange rates. During 1975-1995, the correlation coefficient between the interest rate on six-month U.S. Treasury bills and the real exchange rate was 0.70. During the 1980s, when real rates were particularly volatile, the correlation coefficient was 0.72, which supports the argument that rising real rates tend to attract an influx of capital and affect the exchange rate.

A rising dollar may hurt the competitiveness of U.S. exports and put downward pressure on the prices of farm exports. From 1975 to 1995, the correlation between the real interest rate on six-month Treasury bills and real agricultural exports was about -0.44 and the correlation between the real interst rate and real crop prices was about -0.14. The theoretical and empirical work by Schuh (1974), Chambers and Just (1979, 1981), and others suggests that the increased real exchange rate will depress agricultural commodity prices. There is no consensus about the magnitude of the price effects, but there is general agreement that a stronger dollar exerts downward pressure on the price of commodities such as corn and soybeans. This negative correlation between real interest rates and commodity prices creates another link between interest rate risk and credit risk.[16] When lenders pass their interest rate risk on to farmers using adjustable-rate loans, they potentially take on additional default risk because of this linkage. Therefore, the pricing of adjustable-rate loans should reflect the level of interest rate risk and expected default risk. Since the risks are correlated, a comprehensive risk management strategy is more likely to jointly address these risks.

Credit Risk and Underwriting Standards

Credit risk evolves from several sources: variability of borrower earnings and an unexpected reduction in the ability of the borrower to repay, decline of asset values and the associated loss of borrower capital and collateral securing the loan, and uncertainty about the willingness of the borrower to repay the loan. Credit risk is clearly a major risk consideration in agricultural financing institutions. In a recent survey of agricultural bankers, the relative importance of various sources of risk were identified (Patrick and Ullerich 1996). Among the bankers who responded, the most important sources of risk (in decreasing order of importance) are: (1) credit availability (the borrower's financial condition), (2) changes in family relationships, (3) crop and livestock price variability, (4) crop yield variability, and (5) injury, illness, or death of the operator/borrower. This set of responses points out the wide variety of factors thought to contribute most to credit risk in the portfolios of agricultural bankers.

The relative ranking of responses to these risks, as identified by agricultural bankers, is similarly interesting and diverse in the areas of production, marketing, and financing. The most important production responses (in decreasing order of importance are: (1) being a low-cost producer, (2) diversification of farm enterprises, and (3) having backup management and labor. The marketing responses are: (1) participation in the government farm program, (2) forward contracting the selling price, and (3) establishing minimum price contracts. The set of financial responses include: (1) management of the debt/leverage position, (2) off-farm employment, and (3) carrying liability insurance. These responses indicate the risk environment that agricultural lenders must incorporate into their credit risk management databases and strategies.

From an aggregate perspective, two underlying issues are important to consider. First, credit risk derives largely from the factors that destabilize farm incomes and asset values—factors such as volatile interest rates, commodity prices, and exchange rates. Second, credit risk is a portfolio management problem, rather than an individual borrower evaluation problem. From the above list of agricultural banker responses, a primary source of credit risk is crop and livestock commodity price variability. That variability derives from supply shocks (which affect the level of output) and from correlated commodity market conditions (which are influenced both directly and indirectly by the level of domestic and foreign demand and, therefore, exchange rates).

It is the instability of farm incomes and farm asset values (particularly, farm real estate), coupled with the lack of a uniform set of financial standards, that has historically led to reduced access of agriculture to national capital markets. Thus, the problem of credit risk in agriculture can be recast in terms of the need to achieve greater uniformity in the preparation and reporting of financial performance information, and the need to attain greater conformity in the use of underwriting standards by agricultural lending institutions.

The need to achieve greater uniformity in financial reporting in order to gain better access to capital markets has been recognized by the Farm Financial Standards Taskforce (Farm Financial Standards Taskforce 1991; Farm Financial Standards Council 1995). The FFSTF cites, "the lack of standardization in farm financial analysis caused problems in understanding and using data for decisions, and [it] was often cited as a substantial barrier to the accessibility of funds from capital markets" (p.1). Further, the National Commission on Agricultural Finance claims that without nationwide standardization, the agriculture industry will have difficulty attracting credit, producers will have difficulty analyzing their operations, and agricultural producers will probably pay a premium for borrowed funds. In this context, the FFSTF identified the strategic goal of establishing universally acceptable financial guidelines for U.S. production agriculture. The major components of that strategy are to identify financial ratios that are common to all parts of the country, to identify standard methods for calculating those ratios, to formulate standard financial statement formats that may be used by all farm lending institutions, and to recommend guidelines for farm financial software (Farm Financial Standards Taskforce 1991, p. 2).

The framework for the FFSTF recommendations is the Financial Accounting Standards Board (FASB) Statement of Financial Accounting Concepts. The framework incoroporates concepts such as valuation methods, determining the value of farm production and net income, accounting for family living withdrawals and income taxes, and depreciation expense. Five financial "criteria" are also recommended by the FFSTF as the basis for analyzing farm financial position and performance: liquidity, solvency, profitability, repayment capacity, and financial efficiency. Specific financial measures (ratios and nonratio indicators) are recommended for each of the five criteria.

The need to attain greater conformity in the use of underwriting standards by agricultural lending institutions has been illustrated by the underwriting standards of the Federal Agricultural Mortgage Corporation (Farmer Mac).[17] A primary objective for Farmer Mac is to increase the availability of long-term credit to farmers and ranchers at stable interest rates. In order to sell farm real estate mortgages into the secondary market, Farmer Mac has established seven key standards for underwriting loans.[18] To guarantee mortgage pool diversity, Farmer Mac employs a risk-based simulation model to determine actuarially sound guarantee fees on structured pools of mortgages. The resulting pools reflect risk-based pricing, which combines the effects of credit risk and pool diversity. When taken together, the characteristics of pool diversity, risk-pricing of loans, and loan monitoring by the originating banks provide effective credit risk management.

As we consider the direction of agricultural financing into the next century, the role of securitization is expected to increase.[19] In this future environment, the ability of financial institutions to remain as portfolio lenders will be challenged by market risks (e.g., interest rate variability) and credit risks. Securitization provides a means of controlling those risks and the costs associated with extending credit to agriculture.

Notes

1. The U.S. (total) trade-to-income ratio is the sum of U.S. exports plus imports divided by U.S. gross domestic product (GDP). The U.S. agriculture trade-to-income ratio is the sum of agricultural commodity exports plus imports divided by farm sector GDP.

2. Fukao finds that the pace of international financial activity has both expanded and undergone significant structural change in recent years. Expansion of financial transactions has been driven by: a larger pool of financial assets that can be held and traded across national borders, the progressive removal of capital controls, and the increased ability to exchange assets into other currencies. Fukao finds that the growth of markets in derivative assets has grown at a spectacular rate in recent years, and that this has reinforced the trend toward growth in institutional investing.

3. Engel and Hakkio suggest that high volatility of exchange rates can alter investor decisions and long-term capital flows may be affected and firms may be reluctant to engage in international trade. If these events spill over into the real economy, or inhibit the smooth functioning of the financial system, monetary policy may also become less effective.

4. In the long run, the money supply is not expected to affect the level of real economic activity, as prices, exchange rates, inflation expectations, and real interest rates are expected to converge to their "fundamental levels," based on the country's productive capacity and consumer preferences.

5. Statistical analysis of the relationship between the slope of the yield curve and measures of economic activity suggest a significant lag between the tightening of monetary conditions and the effects of that tightening on real GDP (International Monetary Fund). In the United States, there is about a five-quarter lag between changes in the yield curve and changes in real GDP. Similarly, the lag between yield curve changes and the inflation rate is about fourteen quarters in length (IMF, p. 49).

6. For similar reasons, there was also quite dramatic growth in the size of the U.S. national debt (from 37 percent of GDP in 1980 to 63 percent in 1994). This compares with increases ranging from 41 percent to 70 percent in the major industrialized countries during the same period.

7. If additional savings offset fiscal deficits, budget deficits would have the same effect as additional taxes. The degree to which tax cuts are offset by private saving is a matter of substantial debate (see Bernheim and the discussants of his paper).

8. The connection between fiscal policy and exchange rates has been tested empirically, but with mixed results. An IMF study of deficit reduction indicates that large deficit reductions through spending cuts are associated with stronger exchange rates (International Monetary Fund).

9. The 1996 FAPRI report argues that several years of market signals will be needed for crop producer decisions to be sufficiently affected to observe an aggregate supply response to the new demand conditions (p. 2).

10. FAIR shifts commodity-price risks from the government to farmers in exchange for a seven-year fixed payment and the understanding that most grain subsidies will end. The commodity-price risk has been shifted back to the farmer to be managed on an individual basis.

11. The "savings and loan" crisis of the 1980s focused the attention of U.S. regulators on the interest rate exposure of depository institutions. The Federal Deposit Insurance Corporation Improvement Act of 1991 (FICIA) now requires that regulators develop procedures such that banks meet capital standards that reflect their level of interest rate risk and insulate depositors from those risks to a greater extent. Regulators have attempted to arrive at a simple standardized model of interest rate risk, but their model was found to have some fundamental flaws (Hanweck and Shull).

12. This problem has been termed "moral hazard" in the finance literature.

13. The FCSIC currently assesses a statutory risk premium, which is 0.25 percent for nonperforming loans in comparison with 0.15 percent for performing loans.

14. Pederson finds that interest rate risk underlies an important set of equilibrium responses at agricultural banks. Bank investments and loans are found to be linked to sectoral rates of loan default and the variability of market interest rates. Variable-rate lending is found to have important own-sector and cross-sector effects in bank portfolio allocation decisions.

15. The percentage of floating-rate loans at commercial banks has declined somewhat during 1994-1996. This decline may be attributed to several factors, including the decline of market rates (and the desire of borrowers to fix their rates at the lower levels) and an increase in the volume of real estate loans provided on relatively short maturities at fixed rates.

16. This additional default risk stems from the farmer absorbing the risk from covariance of commodity prices with interest rates. The correlation between interest rates and the lagged commodity price is found to be relatively strong at about -0.64 during 1980-1995.

17. Farmer Mac was created by the Agricultural Credit Act of 1987 to oversee the development of a secondary market for farm real estate and rural housing loans. Farmer Mac operates as an independent entity within the Farm Credit System. It exists as a government-sponsored enterprise with a federal charter and private ownership. It has access to financial markets through the sale of mortgage-backed debt securities, which have agency status.

18. The seven key standards include creditworthiness of the borrowers, current financial statements, a debt-to-asset ratio test, earnings and liquidity tests, loan-to-appraised-value and cash-flow coverage tests, acreage and annual receipts minimum requirements, and loan terms and conditions. In addition, certain individual loan size limits apply (Barry et al.).

19. Securitization is the pooling and repackaging of similar loans into marketable securities that can be sold to investors. It provides a process for improving the liquidity of assets and capital-to-asset ratios while increasing earnings.

References

Backus, D. K., P. J. Kehoe, and F. E. Kydland. 1994. "Dynamics of the Trade Balance and the Terms of Trade: The J-Curve?" *The American Economic Review* 84: 84-103.

_____. 1993. "International Business Cycles: Theory and Evidence," *Quarterly Review*, Federal Reserve Bank of Minneapolis, Fall: 14-28.

Barry, P. J., B. J. Sherrick, and P. N. Ellinger. 1997. "Farmer Mac's New Environment: Key Issues and Performance Factors." The Center for Farm and Rural Business Finance, University of Illinois.

Battan, D. S., and M. T. Belongia. 1984. "The Recent Decline in Agricultural Exports: Is the Exchange Rate the Culprit?" *Review*, Federal Reserve Bank of St. Louis, 66: 5-14.

Becketti, S. 1993. "Are Derivatives Too Risky for Banks?" *Economic Review*, Federal Reserve Bank of Kansas City, Third Quarter: 27-42.

Belous, R. S. 1992. "Shifting Global Markets: An Introduction," in *Global Capital Markets in the New World Order*. Washington, DC: National Planning Association.

Bernheim, D. B. 1987. "Ricardian Equivalence: An Evaluation of Theory and Evidence," in *National Bureau of Economic Research Macroeconomics Annual 1987*, Stanley Fisher (ed.), Cambridge, MA: MIT Press, pp. 262-315.

Bowden, R. J., and V. L. Martin. 1995. "International Business Cycles and Financial Integration." *The Review of Econ. and Statistics* 77: 305-320.

Bredahl, M. E., W. H. Meyers, and K. J. Collins. "The Elasticity of Foreign Demand for U.S. Agricultural Products: The Importance of the Price Transmission Elasticity," *American Journal of Agricultural Economics* 60: 58-63.

Budzeika, G. 1991. *Determinants of the Growth of Foreign Banking Assets in the United States*, Federal Reserve Bank of New York, Research Paper No. 9112, May.

Chambers, R. 1988. "An Overview of Exchange Rates and Macroeconomic Effects on Agriculture,." in *Macroeconomics, Agriculture, and Exchange Rates*, Philip Paarlberg and Robert Chambers (eds.). Westview Press.

Chambers, R., and R. Just. 1979. "A Critique of Exchange Rate Treatment in Agricultural Trade Models." *American Journal of Agricultural Economics* 61: 249-257.

_____. 1981. "Effects of Exchange Rate Changes in U.S. Agriculture: A Dynamic Analysis," *American Journal of Agricultural Economics* 63(1981):32-46.

Coyle, W. T. 1996. "APEC: Absorbing U.S. Ag Exports," *Agricultural Outlook*, Economic Research Service, U.S. Department of Agriculture, Washington, DC. September, pp. 24-29.

Devadoss, S., and S. Chaudhary. 1994. "Effects of Fiscal Policies on U.S. Agriculture," *Applied Economics* 26: 991-997.

Dornbush, R. 1981. "Expectations and Exchange Rate Dynamics," *Journal of Political Economy* 84: 1161-76.

Economic Report of the President. 1996. Washington, DC: U.S. Government Printing Office.

Engel, C., and C. S. Hakkio. 1993. "Exchange Rate Regimes and Volatility," *Economic Review*, Federal Reserve Bank of Kansas City, Third Quarter: 43-58.

Evans, P. 1986. "Is the Dollar High Because of Large Budget Deficits?" *Journal of Monetary Economics*: 27-49.

Farm Financial Standards Council. 1995. *Financial Guidelines for Agricultural Producers*, Naperville, IL.

Farm Financial Standards Taskforce. 1991. *Recommendations of the Farm Financial Standards Taskforce: Financial Guidelines for Agricultural Producers*, Naperville, IL.

Feldstein, M. 1986. "The Budget Deficit and the Dollar," in *National Bureau of Economic Research Macroeconomics Annual 1986*, Stanley Fisher (ed.). Cambridge, MA: MIT Press, pp. 355-92.

_____. 1995. "Lower Deficits, Lower Dollar," *Wall Street Journal*, May 16.

Food and Agricultural Policy Research Institute. 1996. *FAPRI 1996 International Agricultural Outlook*, Staff Report #2-96, Iowa State University, Ames, IA, September.

Frankel, A. B., and P. B. Morgan. 1992. "Deregulation and Competition in Japanese Banking," *Federal Reserve Bulletin*, Board of Governors of the Federal Reserve System, 78:579-593.

Fukao, M. 1993. *International Integration of Financial Markets and the Cost of Capital*, Working Paper No. 128 (OECD/GD (93)62), Organization for Economic Cooperation and Development, Paris.

Furlong, F., and R. Ingenito. 1996. "Commodity Prices and Inflation," *Economic Review*, The Federal Reserve Bank of San Francisco, 2: 27-47.

Gertler, M., and R. G. Hubbard. 1988. "Financial Factors in Business Fluctuations," in *Financial Market Volatility*, Federal Reserve Bank of Kansas City.

Grabbe, J. Orlin. 1986. *International Financial Markets*. New York, NY: Elsevier Press.

Grassman, S. 1980. "Long-term Trends in Openness of National Economies," *Oxford Economic Papers*, 32: 123-133.

Hakkio, C. S. 1996. "The Effects of Budget Deficit Reduction on the Exchange Rate," *Economic Review*, Federal Reserve Bank of Kansas City, 81:21-38.

Hanweek, G. A., and B. Shull. 1996. Interest Rate Volatility: Understanding, Analyzing, and Managing Interest Rate Risk and Risk-Based Capital. Chicago.

International Monetary Fund. 1996. *World Economic Outlook.* Washington, DC, October.

Kaufman, H. 1994. "Structural Changes in the Financial Markets: Economic and Policy Significance," *Economic Review,* Federal Reserve Bank of Kansas City, Second Quarter: 5-16.

_____. 1992. "Opportunities and Challenges in Global Capital Markets" in *Global Capital Markets in the New World Order.* Washington, DC: National Planning Association.

Keynes, J. M. 1936. *The General Theory of Employment, Interest, and Money.* New York: MacMillan.

Kenen, P. 1985. "Macroeconomic Theory and Policy: How the Closed Economy Was Opened," in *Handbook of International Economics: Vol. 2,* R. W. Jones and P. B. Kenen (eds.), Amsterdam: North-Holland.

Kydland, F., and E. Prescott. 1982. "Time To Build and Aggregate Functions," *Econometrica* 40: 1345-70.

Meade, J. E. 1951. *The Balance of Payments.* London: Oxford University Press.

Melvin, F. 1992. *The Economics of Money, Banking, and Financial Markets.* Third Edition. New York: Harper Collins College Publishers.

Mishkin, F. 1995. "Symposium on the Monetary Transmission Mechanism," *Journal of Economic Perspectives* 9 (Fall): 3-10.

Mussa, M., M. Goldstein, P. B. Clark, D. Mathieson, and T. Bayoumi. 1994. *Improving the International Monetary System, Constraints and Possibilities,* International Monetary Fund, Washington, DC, December.

Nolle, D. E. 1994. *Are Foreign Banks Out-Competing U.S. Banks in the U.S. Market?* Economic and Policy Analysis Working Paper 94-5, U.S. Comptroller of the Currency, Washington, DC, May.

Patrick, G. F., and S. Ullerich. 1996. "Information Sources and Risk Attitudes of Large-Scale Farmers, Farm Managers, and Agricultural Bankers," *Agribusiness* 12: 461-471.

Pederson, G. D. 1992. "Agricultural Bank Portfolio Adjustments to Risk," *American Journal of Agricultural Economics* 74: 672-681.

Roe, T., M. Shane, and D. H. Vo. 1986. *Price Responsiveness of World Grain Markets: The Influence of Government Intervention on Import Price Elasticity.* Technical Bulletin No. 1720, Economic Research Service, U.S. Department of Agriculture, Washington, DC.

Rose, Peter S. 1994. *Money and Capital Markets.* Boston: Irwin Co.

Rosensweig, M., and E. Tallman. 1993. "Fiscal Policy and Trade Adjustment: Are the Deficits Really Twins," *Economic Inquiry* 32: 580-94.

Schuh, G. E. 1974. "The Exchange Rate and US Agriculture," *American Journal of Agricultural Economics* 56: 1-13.

Shane, Matthew. 1990. *Exchange Rates and U.S. Agricultural Trade.* Agriculture Information Bulletin No. 585. Economic Research Service, U.S. Department of Agriculture, Washington, DC.

Shiller, R. J. 1988. "Causes of Changing Financial Market Volatility," in *Financial Market Volatility,* Federal Reserve Bank of Kansas City.

Stallings, D. 1992. *Exchange Rates.* Computer file #88021, Economic Research Service, U.S. Department of Agriculture, Washington, DC, February.

Stamoulis, K. G., and G. C. Rausser. 1988. "Overshooting of Agricultural Prices," in *Macroeconomics, Agriculture, and Exchange Rates*, Philip Paarlberg and Robert Chambers (eds.). Westview Press,.

Teigen, L. D., and M. Shane. 1995. *The Federal Budget and U.S. Competitiveness in World Markets: An Overview*. Staff Paper No. 9521, Economic Research Service, U.S. Department of Agriculture, Washington, DC.

U.S. Bureau of the Census. 1996. *Statistical Abstract of the United States: 1996*. 116th Edition, Washington, DC.

U.S. Department of Agriculture, Economic Research Service. Various years. "Indexes of Real Trade-Weighted Dollar Exchange Rates," *Agricultural Outlook*, Washington, DC.

U.S. Government Accounting Office. 1996. *Farm Credit System: Analysis and Comment on Possible New Insurance Corporation Powers*, GAO/GGD-96-144, Washington, DC, August.

Wagster, J., K. Cooper and J. Kolari. 1994. *The Consequences of Regulatory Discretion in Implementing the Basle Accord: Capital Market Evidence*. Paper presented at the Eastern Finance Association meetings, April.

Wall Street Journal. 1997 "Southeast Asian Currencies Fall Again," August 13.

Walraven, N. A. and D. Carson. 1996. *Agricultural Finance Databook*. Federal Reserve Board of Governors, Washington, DC.

Ziengs, D. 1996. Rabobank Nederland, Communication, New York, NY, December.

Zimmer, G. C., and R. N. McCauley. 1991. "Bank Cost of Capital and International Competition," *Quarterly Review*, Federal Reserve Bank of New York, Winter: 33-59.

2

Changes in the Farm Sector

David H. Harrington, Robert A. Hoppe, R. Neal Peterson,
David Banker, H. Frederick Gale, Jr.

The distribution of farms and farm production by size, tenure, and legal organization has changed—along with the composition of farm household income, technology and business arrangements (contracting), and the role of government in the agricultural sector—to significantly alter the performance and financial characteristics of farms. As the structure of agriculture continues to change, such topics as large-scale and corporate farming, contract hog production, bST in dairy production, the declining number of farms, and the survival of family farms punctuate the debate.

The structure and organization of the farm sector are steadily evolving, driven by changes in production technology, off-farm opportunities, and the organization of markets and linked industries. Structure of the farm sector is frequently defined to encompass:

- The numbers and sizes of farms,
- Concentration of farm production,
- Ownership and control of productive farm resources including land,
- Enterprise and regional specialization of production,
- Legal forms of business organization,
- Contractual linkages with other farms or nonfarm businesses, and
- Characteristics of farm operator households.

Structural characteristics, along with institutional arrangements, are of equal importance with traditional economic relationships in influencing the financing

The views expressed are those of the authors and do not necessarily reflect the position of the Department of Agriculture or the current administration.

and performance of the agricultural sector. We concentrate on interpreting how recent changes in the structure and institutional setting of the agricultural sector —particularly the structure of farm household income—may influence the financing of agriculture over the next generation.

Developments in the Sector: Industrialization, Market Orientation, and Dual Structure

Many familiar structural descriptors—such as numbers of farms and average sizes—convey little about the diversity of the farming sector. We briefly review recent trends and present new evidence on the distribution of farms, sales, and assets by (constant dollar) sales class over time to provide additional insight into recent changes.

Structure of the Sector in Brief

Farm numbers, as measured by the Bureau of the Census, declined by 64 percent from 5,388,000 in 1950 to 1,925,000 in 1992. At the same time, land in farms declined by only 19 percent, resulting in a 127-percent increase in acres per farm from 216 to 491. The value of products sold measured in constant (1992) dollars nearly tripled from $54.8 billion in 1950 to $162.6 billion in 1992.

Distributions of farms, land, value of sales, and value of assets from the 1992 Census of Agriculture data are presented in Table 2.1, along with averages per farm. Selected financial characteristics, by sales class, from the 1993 Farm Costs and Returns Survey (FCRS) are shown in Table 2.2. Both tables highlight the diversity of the farm sector and illustrate that generalizations from sector averages—whether about sizes, incomes, or importance of segments of the farm sector—are misleading. Wide variations and strong correlations, positive or negative, among many physical and financial measures make generalizations from averages hazardous.

Constant Dollar Sales Classes. Between 1950 and 1992, the relative share of farms and value of products sold accounted for by farms in different size categories changed significantly (Table 2.3). Comparisons of farms by sales class have traditionally suffered from the fact that inflation causes farms to graduate to higher sales classes, even without a change in production. To overcome this, we use a method of adjusting to constant (1992) dollars (Peterson 1998). In 1950, farms with sales of less than $40,000 accounted for 96 percent of all farms and 63 percent of the value of products sold. By 1992, farms in this category represented 70 percent of all farms but only 8 percent of the value of products sold. In 1992, 63 percent of the value of products sold wwas accounted for by the 7 percent of farms with sales of $250,000 or more. Total value of land and buildings in 1992 was more evenly distributed; farms with sales of less than $40,000 accounted for 35 percent of the total, and farms with sales of $250,000 or more accounted for 30 percent.

TABLE 2.1 Distribution of Farms, Land, Market Value of Products Sold, and the Value of Major Physical Assets, 1992

Item	Farms	Land	Value of Sales[a]	Value of Assets[b]	Land	Value of Sales[a]	Value of Assets[b]
		Percent of U.S. total			*Acres per farm*	*Dollars per farm*	
All farms	100.0	100.0	100.0	100.0	491	84,459	405,661
Farms by production specialty:							
Crops	44.8	37.7	45.3	54.0	413	85,329	488,886
Cash grains	21.0	25.9	20.2	30.1	605	81,070	579,788
Field crops	13.0	7.8	9.0	11.2	293	58,669	348,003
Cotton	1.1	2.0	2.8	2.6	939	223,847	992,871
Tobacco	4.7	1.2	1.6	2.0	126	29,067	173,316
Sugarcane/beets	0.2	0.4	1.0	0.8	959	400,762	1,411,275
Irish potatoes	0.2	0.4	1.3	0.7	793	448,086	1,218,534
Other field crops	6.8	3.7	2.3	5.1	270	28,761	305,251
Vegetables, melons	1.5	0.8	3.7	2.3	256	205,911	607,905
Fruit and tree nuts	4.6	1.0	5.6	5.8	108	101,135	506,970
Horticultural specialities	2.1	0.2	4.7	2.0	59	191,000	383,722
General crop farms	2.5	2.0	2.1	2.6	387	69,626	421,862
Livestock	55.2	62.3	54.7	46.1	554	83,737	338,803
Livestock, except dairy, poultry	42.0	53.9	31.0	33.6	630	62,341	324,943
Beef cattle feedlots	3.0	2.9	12.7	2.9	471	359,883	388,701
Beef cattle, except feedlots	31.8	45.0	10.8	25.0	695	28,611	318,605
Hogs	4.4	1.9	5.6	3.2	208	107,900	294,934
Sheep and goats	1.5	2.4	0.4	1.1	801	20,478	287,882
Other livestock	1.3	1.7	1.5	1.6	633	97,140	477,544
Dairy farms	5.9	4.0	12.3	6.8	336	176,428	470,769
Poultry and eggs	1.8	0.5	9.7	1.7	138	450,579	379,673
Broiler, fryer, roaster chickens	0.9	0.3	4.7	0.8	137	421,954	348,938
Chicken eggs	0.6	0.1	2.4	0.5	113	361,414	382,564
Turkeys and eggs	0.2	0.1	1.7	0.2	251	811,068	559,594
Poultry hatcheries	0.0	0.0	0.8	0.1	103	3,098,361	947,555
Other poultry/eggs	0.1	0.0	0.1	0.1	100	81,001	252,179
Animal specialties	4.2	1.0	1.0	2.3	118	20,595	225,307
General livestock farms	1.3	2.8	0.7	1.6	1,077	43,821	499,978
Farms by value of sales:							
$9,999 or less	47.1	13.1	1.9	18.7	136	3,357	161,305
$10,000-$39,999	22.7	14.8	5.6	15.2	321	20,853	272,484
$40,000-$99,999	12.9	17.7	10.0	15.2	672	65,279	477,411
$100,000-$249,999	10.8	24.1	20.1	21.3	1,094	156,959	797,817
$250,000-$499,999	4.1	13.8	16.6	13.0	1,666	342,653	1,288,162
$500,000-$999,999	1.6	8.5	12.9	7.8	2,598	675,380	1,961,256
$1,000,000 or more	0.8	8.0	33.0	8.8	4,751	3,377,155	4,321,397

[a] Market value of agricultural products sold.
[b] Value of land, buildings, machinery and equipment.
Source: Economic Research Service, USDA, from the 1992 Census of Agriculture.

TABLE 2.2 Selected Farm Business Financial Characteristics, by Sales Class, 1993

Item	Noncommercial	Commercial					All farms
		Small	Lower medium	Upper medium	Large	Superlarge	
Number of farms	1,514,476	210,478	222,645	70,300	30,575	14,825	2,063,300
				Dollars per farm			
Gross cash income	11,922	76,576	155,124	326,372	571,882	2,226,139	68,891
Livestock sales	5,476	32,826	64,437	123,236	231,956	1,128,005	30,062
Crop sales	2,979	25,648	60,862	144,766	252,146	928,293	26,709
Government payments	1,245	7,409	14,482	25,902	28,019	32,153	4,761
Other farm-related income	2,222	10,692	15,343	32,467	59,761	137,688	7,359
Cash expenses	12,730	62,604	119,564	250,806	462,697	1,829,946	57,182
Net cash farm income	*-808	13,972	35,560	75,566	109,186	396,193	11,709
Net farm income	*1,105	8,709	25,700	50,378	95,287	461,605	10,918
Farm assets	261,606	488,497	656,507	1,079,544	1,497,860	4,013,647	400,511
Liabilities	17,359	74,791	116,519	208,359	316,709	893,219	51,154
Equity	244,247	413,706	539,989	871,185	1,181,151	3,120,428	349,356
				Percent			
Debt/asset ratio	6.6	15.3	17.8	19.3	21.1	22.3	12.8
				Dollars per farm			
Capital investments	4,337	11,958	20,133	36,237	62,551	133,727	9,698

* RSE greater than 25 percent, but no more than 50 percent.
Source: Economic Research Service, USDA, compiled from the 1993 Farm Costs and Returns Survey.

TABLE 2.3 Distribution of Farms, Market Value of Products Sold, and the Value of Land and Buildings, Selected Years, 1950-1992

Item	1950	1969	1992	1950	1969	1992	1950	1969	1992
					Percent				
All farms	100.0	100.0	100.0	100.0	100.0	100.0	100.0	100.0	100.0
Farms by real value products sold:[a]									
$9,999 or less	73.5	48.0	47.1	20.9	6.4	1.9	na	na	19.3
$10,000-$39,999	22.5	28.9	22.7	41.8	15.8	5.6	na	na	15.3
$40,000-$99,999	3.1	15.6	12.9	18.0	24.9	10.0	na	na	14.9
$100,000-$250,000	0.7	5.8	10.8	9.6	21.6	20.1	na	na	20.9
$250,000-$499,999	0.1	1.2	4.1	4.1	10.1	16.6	na	na	12.9
$500,000 or more	0.1	0.6	2.4	5.7	21.2	45.9	na	na	17.0
Farms by product specialization:[b]									
Cash grain	11.6	na	21.0	14.5	na	20.2	20.9	na	29.6
Cotton	16.4	na	1.1	10.2	na	2.8	8.1	na	2.5
Other field crop	11.0	na	11.9	6.8	na	6.2	5.0	na	8.3
Vegetable	1.3	na	1.5	2.3	na	3.7	1.5	na	2.3
Fruit & Nut	2.2	na	4.6	3.4	na	5.6	4.1	na	6.1
Poultry	4.7	na	1.8	5.5	na	9.7	2.8	na	1.6
Dairy	16.2	na	5.9	15.4	na	12.3	12.8	na	6.2
Other livestock	21.7	na	46.2	28.9	na	32.0	30.3	na	37.3
All other	14.7	na	5.9	13.0	na	7.4	14.5	na	6.2
Farms by census region:									
New England	1.9	1.1	1.2	2.4	1.5	1.0	1.6	0.9	1.3
Middle Atlantic	5.5	4.5	4.5	6.3	4.7	4.1	4.6	3.5	4.3
East North Central	16.4	18.8	16.9	18.2	16.6	14.9	19.5	17.7	16.0
West North Central	18.2	23.4	23.4	25.6	27.5	26.7	25.3	24.1	22.3
South Atlantic	17.8	13.6	11.6	9.6	10.7	11.6	9.5	9.9	11.1
East South Central	16.9	14.4	12.2	6.2	6.0	5.7	6.9	7.0	6.6
West South Central	14.5	14.6	16.5	12.9	12.6	13.1	14.1	16.7	14.6
Mountain	3.6	4.4	6.1	7.3	8.4	8.3	7.3	8.4	10.1
Pacific	5.1	5.3	7.6	11.6	12.1	14.6	11.1	11.7	13.6

[a] Sales class boundaries, farm numbers and value of sales are adjusted to constant (1992) dollars. [b] All farms in 1992, all commercial farms in 1950.

Source: Economic Research Service, USDA, from the Census of Agriculture.

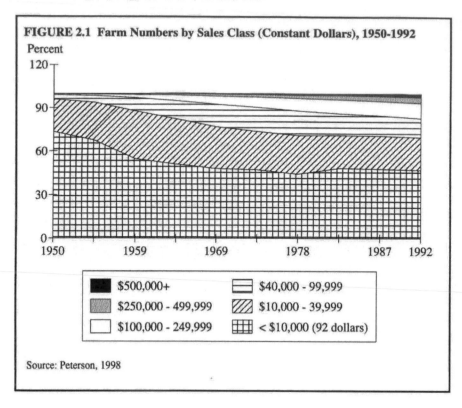

FIGURE 2.1 Farm Numbers by Sales Class (Constant Dollars), 1950-1992

Legend:
- $500,000+
- $250,000 - 499,999
- $100,000 - 249,999
- $40,000 - 99,999
- $10,000 - 39,999
- < $10,000 (92 dollars)

Source: Peterson, 1998

The share of farms with sales less than $10,000 (1992 dollars) has remained relatively constant since the 1969 Census, reaching a low of 44 percent in 1974 but otherwise varying between 47 and 48 percent. The share of farms with sales of $10,000-$39,999 first expanded and then contracted as the larger sales classes expanded faster (Figure 2.1). The share of farms with sales valued at $10,000 to $39,999 reached a peak of 33 percent in 1959, but declined thereafter and has been relatively constant at 23 percent in the last three census years (1982, 1987, 1992). Farms with sales valued at $40,000-$99,999 peaked at 17 percent of farms in 1978 and declined to 13 percent in 1992. The share of farms in the three largest sales classes has increased steadily from 1950 to 1992.

Production has become increasingly concentrated on farms with sales valued at $250,000 or more (Figure 2.2). While the share of farms with sales valued at $100,000-$249,999 has continued to increase (Figure 2.1), the corresponding share of the value of products sold has decreased in the last three census years (Figure 2.2). The share of assets has, for the most part, increased on farms with sales valued at $250,000 or more since 1950 (Figure 2.3). Nevertheless, over one-third of the value of land and buildings was still accounted for by farms with sales less than $40,000 in 1992 (Table 2.3). This result, combined with contin-

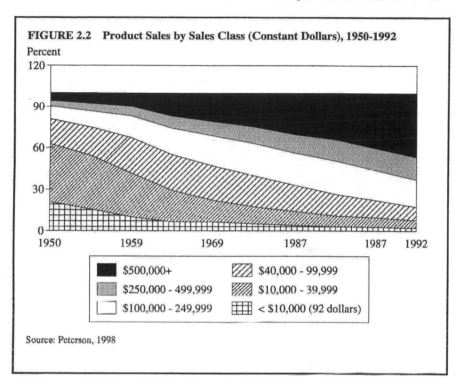

FIGURE 2.2   Product Sales by Sales Class (Constant Dollars), 1950-1992

Percent

Legend:
- $500,000+
- $250,000 - 499,999
- $100,000 - 249,999
- $40,000 - 99,999
- $10,000 - 39,999
- < $10,000 (92 dollars)

Source: Peterson, 1998

ued growth in the share of farms in the three largest sales classes, suggests the continued evolution of a "dualistic" agriculture. Namely, a slowly increasing group of largest farms (sales of $250,000 or more) account for a substantially increasing share of sales and a moderately increasing share of farm assets, while smaller farms (sales of less than $40,000) account for a relatively constant share of farm numbers and assets but a substantially declining share of sales.

The Changing Composition of Income

Perhaps the most striking changes over time have been in the composition of farm income. As with farm structural descriptors, these changes reflect the influence of both market forces and government policy. In recent years, interest has focused on changes in payments from government programs and farmers' growing off-farm income. Measurement of incomes in agriculture, however, requires some clarification of concepts in order to identify and interpret the changes. Moreover, measuring changes in the composition of farmers' income is complicated by the fact that the farm sector can be viewed from three levels: the aggregate sector level, the business level, and the household level (Figure 2.4). Government payments are most logically measured at the sector or business lev-

FIGURE 2.3 Land and Buildings by Sales Class (Constant Dollars), 1950-1992

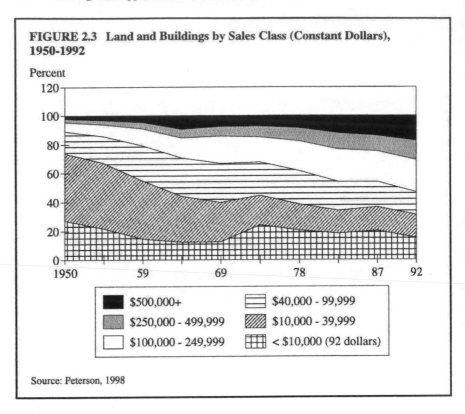

Source: Peterson, 1998

els as components of gross income from farming. Off-farm income is most logically measured at the household level as a component of household income.

The term "farm sector" is often used but seldom defined. The farm sector actually consists of a variety of businesses sharing the income generated by farming. These businesses include farms (whatever their legal organization), nonfarm contractors, and nonfarm share landlords. Share landlords and contractors who also farm are classified as farm businesses in Figure 2.4. These businesses generate gross cash income, pay expenses, and distribute net income to various types of households.

The households associated with sector businesses receive the net income generated by businesses. Operator households are perhaps the most closely connected to farm businesses. As defined in this chapter, operator households are the households of farm operators running farms organized as proprietorships, partnerships, and family corporations. These farms are closely controlled by their operator households. Farms organized as nonfamily corporations or cooperatives are excluded, as are farms with hired managers, because the operators and their households have only limited control of their farms. Households other than those

FIGURE 2.4 Farm Sector Participants

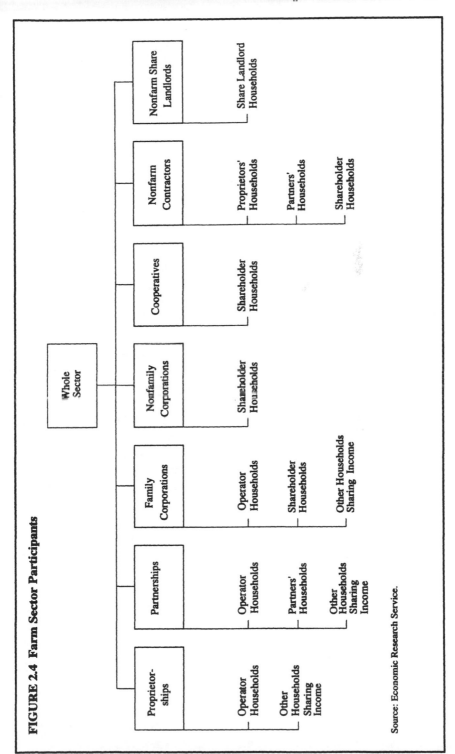

Source: Economic Research Service.

of the operators also share in farm income. For example, operator households may share net income of the farm with the households of partners and shareholders. Even farm proprietorships may share informally. For example, one sibling may operate the family farm, but share the farm's income with other siblings.

Composition of Gross Cash Income. The Economic Research Service (ERS) maintains historical estimates of sector income by source (Figure 2.5). These estimates include all the entities shown at the business level in Figure 2.4 except nonfarm landlords.

Crop and Livestock Marketings. Most gross cash farm income comes from crop and livestock marketings. The relative importance of these items has changed over time due to the increased importance of grain exports. Before 1972, livestock marketings always exceeded crop marketings, often by a large margin (Figure 2.5). Between 1972 and 1974, however, the share of gross cash income from crops increased from 39 percent to 55 percent, and the share from livestock declined from 54 percent to 44 percent. Crop marketings nearly doubled from $26 billion in 1972 to $51 billion in 1974, due to increased export demands. The increase in crop marketings reflected the lower grain inventories in 1971 and 1972 followed by poor world grain harvests in 1972 and 1974 (Cochrane 1993, p. 155). Livestock marketings also increased, but at a slower rate, from $36 billion to $41 billion. Since 1974, the shares of gross cash income from crops and livestock never varied from each other by more than 5 or 6 percentage points.

Government Payments. Government payments peaked at 10.2 percent of gross cash income in 1987 and totaled at least 2 percent of gross cash farm income for two extended periods—1957-1973 and 1982 to the present.[1] The earlier period roughly corresponds with an era of chronic oversupply (Cochrane 1993, p. 139). The beginning of the second period coincides with the end of the export boom of the 1970's and the beginning of the extended farm financial crisis of the 1980's (Cochrane 1993, p. 166). Additionally, high payments occurred in 1993 as a result of high feed grain production, and many farmers received disaster payments in 1993 and 1994 for droughts or floods during 1993 (Perry and Morehart 1994, p. 19).

An inverse relationship has existed historically between the percentage of gross cash income from government payments and the percentage of gross cash income from crop marketings. As one would expect from the purpose of the programs, when the share of income from crops goes down, the share of income from government payments goes up. Examining government payments for the sector as a whole obscures the fact that government payments are more important to some farms than to others. Government payments tend to be highest for larger farms and, not surprisingly, farms specializing in program commodities (Perry and Morehart 1994, p. 19). Other farms may also receive program payments, however, such as livestock farms growing program crops as feed.

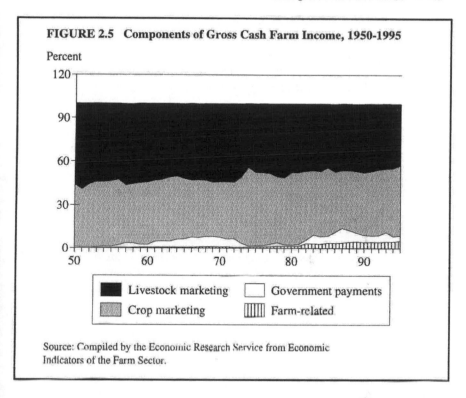

FIGURE 2.5   Components of Gross Cash Farm Income, 1950-1995

Source: Compiled by the Economic Research Service from Economic
Indicators of the Farm Sector.

Other Farm-Related Income. Other farm-related income increased from 0.1 percent of gross cash income in 1950 to 5.3 percent in 1995. A large part of the increase, however, is due to data improvements. As more data became available, more income sources were added. From 1950 to 1963, the category was solely machine hire and custom work. Recreational income was added in 1964, and a category termed "additional sources" was added in 1974. Starting in 1978, forestry product sales were shifted to other farm-related income from crop marketings. Finally, livestock and poultry contract production fees were added in 1982.

Composition of Operator Household Income. Dependence on off-farm income is not a recent phenomenon—just a recently discovered phenomenon. Although there are no consistent data series on operator household income extending back to 1950, various data can be pieced together to give an idea of operator households' sources of income through the years (see acompanying text box). ERS income data indicate that operator households relied on off-farm sources for at least 50 percent of their income at least since the 1960s (Figure 2.6).

Box 2.1 Measuring Operator Household Income

Household income, as used in this chapter, is defined to be consistent with the money income concept used by the Census Bureau. This allows these estimates of operator household income to be compared with total U.S. household income from the Current Population Survey (CPS), conducted by the Census Bureau. Money income includes any income received as cash but excludes income received in-kind. The Census Bureau departs from a strictly cash concept by deducting depreciation as an expense for the self-employed.

Both farm and off-farm sources of income are included when measuring operator household income. Household farm income includes the operator household's share of their farm's cash income less cash expenses and depreciation. Also included as farm income are wages paid by the farm business to household members and net income received by the household from another farm business.

From 1988 onward, the Economic Research Service (ERS) has prepared operator household income estimates using data from the Farm Costs and Returns Survey (FCRS). Information is collected for only one operator and one operator household per farm. For farms with more than one operator, data are collected only for the primary operator and his or her household.

Prior to using FCRS-based estimates, ERS divided net farm income for the sector by the number of farms to estimate farm income per operator household. An estimate of off-farm income per household was then added to farm income per household to derive total household income. There were problems with this procedure, however. First, it was not consistent with CPS (and hence FCRS) methodology. And second, it assumed the operator household received all the income of the farm. Ahearn (1986) devised a procedure to make the sector-based estimates more consistent with the CPS (and FCRS) for the years 1960-1984. The pre-FCRS household income estimates in Figure 2.6 have been adjusted for revisions in the sector accounts since Ahearn published her series. However, the sector-based estimates in Figure 2.6 still assume the operator household receives all the farms' income.

It is difficult to say exactly when farm households began to rely heavily on off-farm income. In fact, operator household dependence on off-farm income in the 1960s is understated in Figure 2.6 because the sector-based estimates of income from farm sources assumed that the operator household received all the income generated by the farm and shared it with no other households, thus overestimating the proportion of income from farm sources. Thus, the share of income from farm sources dropped noticeably when ERS switched to the Farm Costs and Returns Survey (FCRS), which collects data on the share of the farm's income received by the operator household.

FIGURE 2.6 Share of Operator Household Income From Farm-Related and Off-Farm Sources, 1960-1995

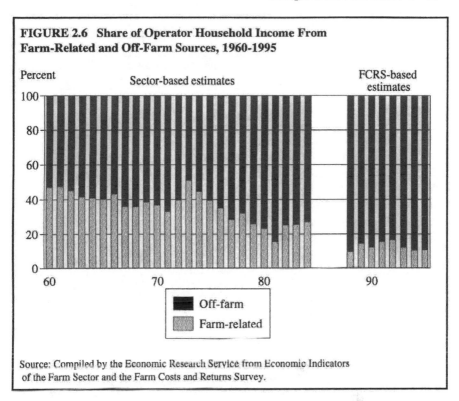

Source: Compiled by the Economic Research Service from Economic Indicators of the Farm Sector and the Farm Costs and Returns Survey.

Although operator household data were not estimated before 1960, data on per capita disposable income of farm residents were provided by the USDA for 1934-1983, including estimates of income from farm and nonfarm sources. The series shows the heavy reliance on off-farm income sources throughout the 1960s, 1970s, and early 1980s (Ahearn 1992, p. 11). Even in the 1930s, 30 to 41 percent of farm residents' personal disposable income came from off-farm sources.

Data from the Census of Agriculture extend even further into the past and corroborate that off-farm income has been important as long ago as the 1930s (Table 2.4). This series shows that one-fourth to one-third of farm operators worked off-farm in the 1930s and early 1940s, generally for fewer than 100 days. By 1954, about 45 percent of operators worked off-farm, only about 7 percentage points less than in 1992. Although the percentage working off-farm has not increased dramatically since 1954, the percentages working at least 200 days off-farm increased from 22 percent in 1954 to 35 percent in 1992, with most of the increase coming before 1969. Since the 1970s, one-third of farm operators have been employed off the farm essentially full time.

TABLE 2.4 Off-Farm Work by Farm Operators, 1930-1992

Census year	Total reporting off-farm work	Operators who reported working off-farm		
		1-99 days	100-199 days	200 plus days
		Percent of all operators		
1930	30.3	18.7	5.2	6.3
1935	30.5	19.3	5.1	6.1
1940	28.7	13.2	6.2	9.3
1945	26.8	8.3	4.2	14.2
1950	38.9	15.6	5.8	17.5
1954	45.0	17.1	6.4	21.5
1959	44.9	15.0	6.2	23.7
1964	47.2	14.2	6.9	26.1
1969	54.2	14.3	8.0	31.9
1974	N.A.	N.A.	N.A.	N.A.
1978	53.3	11.2	8.0	34.1
1982	53.0	10.0	8.4	34.6
1987	53.4	9.6	8.5	35.3
1992	51.6	8.6	8.4	34.6

Source: Census of Agriculture, various years.

Currently, operator household income comes from a variety of sources in addition to the farm. On average, 89 percent of operator households' income came from off-farm sources in 1995 (Figure 2.7), with off-farm wages and salaries being the largest source. Smaller farms are more reliant on off-farm income than are larger farms. About three-fourths of U.S. farm households operated noncommercial farms—which, on average, lost $3,400 farming and pulled down the average share of household income from farming (Table 2.5). Households with commercial farms earned more from farming and received only about 50 percent of their income from off-farm sources. The percentage from off-farm sources declined with farm size within the commercial category.

Average income for all operator households, however, was similar to that for all U.S. households (Hoppe and others 1996). Most groups in Table 2.5 had total household income near or above the average for all U.S. households in 1995. Groups with substantially lower than average income included the young (age 34 or under) and the old (age 65 or older).

Changing Production Arrangements

Traditional family farming, as practiced up to a generation ago, has been steadily giving way to more diverse forms of organization, combinations of enterprises, and sources of income.

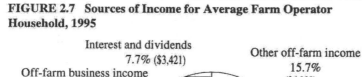

FIGURE 2.7   Sources of Income for Average Farm Operator Household, 1995

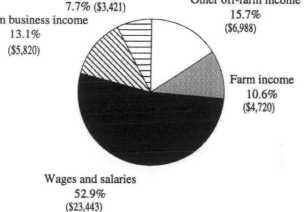

Interest and dividends
7.7% ($3,421)

Off-farm business income
13.1%
($5,820)

Other off-farm income
15.7%
($6,988)

Farm income
10.6%
($4,720)

Wages and salaries
52.9%
($23,443)

Source: Compiled by the Economic Research Service from the 1995 Farm Costs and Returns Survey.

Small Farming Requires Combining Diverse Income Sources. Many farms once thought to have resource and asset bases insufficient to fully employ an operator and farm family can now combine traditional crop and livestock farming with off-farm employment and contract production to greatly enhance incomes. Combining permanent off-farm employment, retirement income, or income from the renting out of assets formerly operated allows small farms to provide comfortable livings for their operator households. Since the 1970s, over one-third of all farms have been operated by a person with a full time off-farm job; an additional 17 percent of farm operators classify themselves as retired (Hoppe and others p.85). Thus, over half of all farms are using this economic strategy. Most of these farms are in the lower sales categories. The average gross cash income per farm (1993) of operators with occupations other than farming was $17,881; for those who are retired, $9,832; and for those whose occupation was farming, $127,151 (Hoppe and others 1996, p. 85).

Large and Superlarge Farming Rely Increasingly on Contracting. Production contracting and vertical integration has become increasingly important in agricultural production. Typically, a processor, feed mill, or other entity in the marketing chain owns and supplies the inputs, the growing commodity, and the processing facilities; and enters into contracts with farmers to grow specified quantities of a commodity, using specified methods, and for delivery at specified times. Contracting first became widespread in the broiler and poultry industries in the 1950s and 1960s. It has more recently become prevalent in hog

TABLE 2.5 Farm Operator Households and Their Income, by Selected Characteristics, 1995

Item	Operator households	Distribution of households	Average total operator household	Compared w/U.S. household income[b]	Share from off-farm sources
	Number	Percent	Dollars	Percent	
Total households	2,036,810	100.0	44,392	98.8	89.4
Operator age:					
34 or under	168,825	8.3	32,506	72.3	93.4
35 to 44	407,345	20.0	47,266	105.2	89.3
45 to 54	476,807	23.4	51,953	115.6	91.6
55 to 64	469,052	23.0	50,421	112.2	87.7
65 or over	514,780	25.3	33,518	74.6	87.2
Operator education:					
Some high school or less	425,612	20.9	30,173	67.1	94.4
Completed high school	819,087	40.2	41,479	92.3	87.3
Some college	443,374	21.8	48,726	108.4	85.8
Competed college	238,561	11.7	55,483	123.5	86.4
Graduate school	110,176	5.4	79,515	176.9	103.4
Operator occupation:					
Farming	903,820	44.4	40,342	89.8	64.8
Something else	797,718	39.2	53,425	118.9	108.9
Retired	335,272	16.5	33,815	75.2	94.9
Type of farm:					
Cash grain	383,554	18.8	48,922	108.9	73.7
Other crops	468,177	23.0	53,476	119.0	79.5
Beef, hogs, or sheep	947,190	46.5	37,605	83.7	108.5
Dairy	121,506	6.0	47,707	106.2	47.8
Other livestock	116,383	5.7	44,695	99.5	109.0
Sales class of farm:					
Noncommercial[c]	1,514,542	74.4	39,814	88.6	108.5
Commercial[d]	522,268	25.6	57,667	128.3	51.1
$50,000 to $249,999	407,851	20.0	40,615	90.4	72.2
$250,000 to $499,999	71,674	3.5	72,307	160.9	40.5
$500,000 or more	42,743	2.1	195,825	435.8	16.0
Legal organization of farm:					
Individual	1,880,516	92.3	42,354	94.2	93.3
Partnership	100,226	4.9	64,387	143.3	68.0
Family corporation	56,067	2.8	76,978	171.3	49.5

[a] Income from off-farm sources can be more than 100 percent of total household income if farm income is negative. [b] Total operator household income divided by U.S. average household income ($44,938).

and beef production. Marketing contracts—wherein the farmer owns the growing commodity but is assured of a specific market and/or price for it—have been used extensively in fruit and vegetable production and now extend to many crop commodities.

Farmers enter into contracts for many reasons, chief among them:

• Improved access to capital,
• Security of markets,
• Improved production efficiency,
• Reduced production and/or price risk.

Similarly, contractors enter into contracts to strengthen their competitive positions by:

• Controlling the quantity, uniformity, and timing of their input supply;
• Expanding and diversifying their operations;
• Better adjusting their own production to market conditions;
• Establishing a known identity in the market for their products;
• Reducing risks from excessive market volatility for their inputs.

Farm Costs and Returns Survey data indicate that while 11 percent of farmers used contracts in 1993, almost one-third of the value of agricultural production was produced under contract: 89 percent in poultry production; 50 percent in fruit/vegetables and dairy;[2] 18 percent in cattle, hogs, and sheep; and 6-12 percent in traditional grains. The use of contracting is heavily associated with larger farms. Among farms with sales of over $500,000, 54 percent of farms use contracts, accounting for 44 percent of their production (Johnson and others 1996, p.7). For farms with sales between $250,000 and $499,999, 40 percent of farms use contracts, accounting for 30 percent of their production. Farms with sales between $50,000 and $249,999 reported 25 percent using contracts, accounting for 22 percent of their production. The smallest sales class, under $50,000, reported 4 percent using contracts, accounting for 11 percent of their production.

Partnerships Are Used Increasingly in Large Farming, Family Corporations in Superlarge Farming. While the overwhelming majority of farms (91 percent, accounting for 66 percent of value of sales) are organized as individual or family proprietorships, FCRS data for 1993 show partnerships and family corporations are important in larger scale farming (Hoppe and others 1996). Farm partnerships, 6 percent of farms and accounting for 16 percent of production, have average sales almost four times as large as those organized as proprietorships. Family corporations have average sales nearly eight times as large as proprietorships and twice as large as partnerships. Corporations involved in farming are overwhelmingly family-held, accounting for only 2.8 percent of farms but 15 percent of production. Nonfamily corporations account for only 0.5 percent of farms and 3 percent of production.

Changing Factor Ownership and Financial Structures

Land ownership/tenure and financial strength of farms are structural characteristics important in agricultural finance decisions. Land tenure has been surprisingly stable over long periods of time, while the financial strength of farmers has varied cyclically with agricultural production and market conditions.

Tenure Arrangements. For reasons including investment, inheritance, and increases in the amount of land one operator can reasonably farm, more people own farmland than operate it. However, the ownership and tenure patterns of U.S. agriculture have remained very stable over time. Noncommercial farms (sales less than $50,000) are predominantly operated by full owners (who account for 56 percent of farms, and 27 percent of the land); part owners, who dominate the commercial sizes from small to superlarge, operate 36 percent of farms and account for 61 percent of land. Full tenants operate only 8 percent of farms and account for 12 percent of land, spread almost evenly across the commercial sizes (Hoppe and others 1996, pp. 6-8). These ratios have barely changed over the last thirty years. However, the age structure of farm operators has changed in recent years. The proportion of farm operators age 65 and over increased by nearly a third to 24 percent between 1982 and 1992. This implies that there may be a large turnover in farm operators, and possibly some changes in tenure, in the coming years.

Financial Structure and Profitability. Table 2.2, adapted from Hoppe and others (1996), summarizes incomes, expenses, assets, debt levels, and equity by sales class for 1993. Noncommercial farms have the lowest net farm incomes and the lowest debt/asset ratios (6.6 percent), yet represent significant accumulations of equity, at $244,200 per farm. Debt/asset ratios among commercial farms increase from 15.3 percent to 22.3 percent as the sales class increases. Net farm income increases from $8,700 to $461,600 and equity increases from $413,700 to over $3.1 million over the commercial size range.

By ERS financial strength categories,[3] 59 percent of all farms were in the favorable category in 1993. Twenty-nine percent, mostly noncommercial, fell in the marginal income category; 6 percent were marginal solvency; and 5 percent were vulnerable. The proportion of farms in the marginal solvency category peaked in 1986 at 11.7 percent, while the proportion of farms in the vulnerable category peaked in 1985 at 10.0 percent. (Morehart 1995).

Developments in Agricultural Markets and Policy

Changing Marketing Arrangements

Marketing arrangements in agriculture are also changing. Traditional livestock, produce, and terminal markets have declined in importance as more of agriculture uses production and marketing contracts. Markets for smaller farms and producers of specialty crops have become, in many cases, niche markets, wherein the producer attempts to exploit special characteristics of the market, based on commodities produced, specifications (e.g., organic production), value-added, seasonality, or location. Markets for major grains, oilseeds, and fibers have become more international with the United States' entry into trade agreements like the North America Free Trade Agreement (NAFTA) and the World Trade Organization (WTO).

Open Markets: Spot, Futures, and Options. In grain and oilseed production, producers still rely heavily on local elevators and cooperatives as their access to the market, but they are making increasing use of forward contracts in the sale of their crops. Futures and options markets are intended to shift the risk of price changes from those who hedge in the market to those who speculate in the market. Although less than 5 percent of grain and oilseed producers traded on futures and options exchanges in 1995, these markets facilitate forward contracting by producers even if the producer does not use them directly. When a producer enters into a contract for future delivery with an elevator or dealer, the elevator typically hedges its position in the futures/options markets in order to minimize its exposure to price risks. This essentially means that the elevator or dealer is using the futures/options markets on the producer's behalf. Marketing contracts in grains and oilseeds were used for over 12 percent of production in 1993 (USDA 1996, p. 8), bringing total use of futures/options markets (including indirect use by elevators/dealers) to as much as 17 percent of production. Direct or indirect reliance of producers on these markets will likely increase as government price supports (loan rates) erode to lower effective levels.

Niche Markets. Niche markets are becoming increasingly important for specialty production and small farming. Examples are organic produce, identity-preserved products such as tofu soybeans, roadside marketing, farmers' markets, pick-your-own commodities, and exotic commodities such as kiwi fruit and Belgian endive. These frequently provide higher returns per acre than traditional farm commodities; but require higher levels of inputs and management and serve markets that are more easily saturated and subject to fads. Data from the 1995 FCRS indicate that 15.3 percent of farms sell some product directly to consumers, and 5.2 percent sell directly to wholesalers and retailers. Direct sales (to consumers) increase to 18 percent of family corporations and 25 percent of nonfamily corporations. Twelve percent of family corporations and 14 percent of nonfamily corporations report selling directly to wholesalers and retailers.

International Markets. With the improvement in access to international markets negotiated under the World Trade Organization (WTO) and the North American Free Trade Agreement (NAFTA), markets for bulk commodities—and value-added products made from them—should expand as these agreements are implemented. There should be increased demand and less volatility in world markets for these goods. However, the commitment of both the United States and its trading partners to freer trade in agriculture does not appear as robust as in some other sectors. Agricultural trade disagreements and retaliatory trade practices still occur, even within the WTO framework. Also, to the extent that former commodity policies and programs in the United States guaranteed minimum prices, set target prices, stimulated U.S. exports, or insulated U.S. markets from international competition, there may be decreases in the prices and increases in the price volatility of those commodities.

The Changing Role of Government

The amount of government assistance and support to the agricultural sector has been declining from peak levels during the farm financial crisis of the mid-1980s and the drought conditions of the early 1990s. With the passage of the Federal Agricultural Improvement and Reform Act of 1996, the role envisioned for government in the agricultural sector has changed dramatically (Nelson and Schertz 1996).

Direct Support. Direct support of agricultural prices and incomes traditionally was accomplished through commodity price support operations (nonrecourse loans) and deficiency payments. Payments for land retired from production for conservation uses were made under the Conservation Reserve Program. Disaster payments were frequently made in cases of natural disasters. With the passage of the 1996 Act and other legislation of the early to mid-1990s, each of these components of government assistance to agriculture is scheduled to decline and possibly end by 2002 (Young and Westcott 1996, pp. 5-10). Total transfer payments to the agricultural sector under direct payment programs have declined from nearly $14 billion in 1993 to $8 billion (forecast) in 1997. Payments are projected to decline to less than $6 billion in 2002, and to end thereafter.

In addition, deficiency payments tied to prices of supported commodities have been replaced by predetermined and declining production flexibility contract (PFC) payments, which though higher than the payments that would have been made under the previous legislation, no longer stabilize farm income. Further, governmental authority to restrict production by requiring acres to be set-aside in order to qualify for direct payments was revoked under the 1996 Act (Nelson and Schertz 1996). The increased flexibility of planting and the reduction of direct support implies that farmers are now much more dependent on the market, and that significant market expansion will have to take place to replace the transfer payments currently scheduled to end in 2002. Although Economic Research Service (USDA) projections indicate minor impacts on aggregate production and price levels, planted acreages, supplies, and prices for formerly supported commodities may become more volatile than in the past.

Market Intervention. In addition to direct payments, interventions in markets for dairy products, sugar, and peanuts have supported prices through direct purchases (dairy products), production quotas and administered prices (peanuts), and tariffs, import quotas, and nonrecourse loans (sugar). The 1996 Act phases out dairy price supports by 2000 and institutes a recourse loan program for butter, cheese, and nonfat dry milk at the equivalent of $9.90 per hundredweight for milk, starting in 2000 (Nelson and Schertz 1996). The peanut program is converted to a "no net cost" program that, theoretically, would reduce transfer payments to peanut producers. The sugar program was continued, but with modifications that reduce government costs.

Trade Intervention. In recent years, funding for the Export Enhancement Program (EEP) has varied between $0.3 and $1.2 billion. The 1996 Act and the Uruguay Round agreement under GATT have reduced the maximum expenditures under EEP to $400 million. With strong international market prospects as of 1997, the United States is unlikely to make full use of authority for EEP expenditures. However, if international markets turn unfavorable, the authority for EEP assistance is very limited and strong trade interventions to protect or restore U.S. markets may be illegal under WTO rules.

Risk Reduction or Transfer. With the elimination of congressional authority to enact disaster payments "off- budget," disaster payments have essentially been eliminated. Farmers, exposed to more yield risk, must bear it themselves or purchase crop insurance. Similarly, with the suspension of deficiency payments and capping of loan rates in the 1996 Act, price volatility may become more pronounced. Producers of formerly supported commodities may use forward contracting or futures and options markets more extensively to manage these price risks. Thus, under the 1996 Act and other recent legislation, farmers appear to be exposed to somewhat higher levels of risk. In recognition of this, the Congress mandated pilot testing of revenue insurance plans wherein an insured producer is covered for shortfalls of crop revenue (commodity price times yield) below a chosen percentage. The pilot tests are still underway in 1997, and there is considerable legislative debate over broadening revenue insurance. Statements by the Secretary of Agriculture in the spring of 1997 indicate that delivery of crop and revenue insurance will be completely turned over to the private sector starting in 1998.

Environmental Protection. The 1996 Act consolidated a wide range of environmental/conservation technical assistance and cost-sharing programs into the Environmental Quality Incentives Program, continued the Conservation Reserve Program, and maintained the requirements for conservation compliance in order to receive PFC payments. Other legislation—such as the Endangered Species Act, the Clean Water Act, the Safe Drinking Water Act, the Federal Insecticide Fungicide and Rodenticide Act, and court decisions enforcing environmental standards and assigning liabilities for cleanup of environmental damages—modifies or restricts some of the property rights of landowners and farmers, and could have financial implications for agricultural lenders as well.

While there has been considerable discussion of the "takings" permitted by these laws—the taking of private property for public use without compensation, in violation of the Fifth Amendment to the Constitution—courts have been reluctant to award compensation, in most cases requiring near total loss of property value. In general, legislation is seldom enacted to compensate a person for activities or opportunities that have been restricted by new laws. To do otherwise, it is argued, would render such basic governmental functions as zoning, health and safety regulations, and pollution control unworkable.

Factors Affecting Agricultural Finance
at the Beginning of the Twenty-First Century

Capital Resources in Agriculture

We focus on three types of capital resources: physical, human, and financial. These types of capital are each affected by different trends and conditions, which sometimes have disparate implications for the financing of agriculture.

Physical Capital. Agricultural technology and productivity can be reasonably assumed to continue at rates experienced since World War II. Total factor productivity in U.S. agriculture grew at an annual rate of 1.95 percent over 1948-1994. In recent years, growth rates have exceeded that long-term average, both in 1979-1989 (2.56 percent) and in 1989-1994 (2.88 percent) (Ahearn and others 1997). These rates reflect rapidly increasing output achieved as intermediate inputs (fertilizers; pesticides; fuels and energy; and feed, seed, and livestock) are substituted for both capital and labor, which declined at 1 to 1.5 percent per year. These robust rates of total factor productivity growth compound the potential problem of insufficient growth in markets for U.S. agricultural products. Agricultural profitability, both domestically and internationally, rises or falls depending on whether or not the long-term growth of demand outstrips the long-term growth of production. In only a few short periods—recognized in retrospect as aberrational—have the prospects of market growth exceeded the prospects for productivity growth. In "normal" postwar periods, the balance between demand growth and supply growth has been maintained by reducing real returns in agriculture, thereby stimulating the exit of resources from farming.

Human Capital: Farm Goals and Decisions. Farm family goals vary from one size of farm to another. While it was once assumed that operation of a full-time farm, without off-farm employment, was the prevailing goal for most farm families, this is no longer the case for the noncommercial and small commercial farms that make up almost 85 percent of farms. These farm families may be pursuing a much wider array of farm and nonfarm opportunities than simply financial profitability of their farms. Other goals might include a desire for a farm lifestyle, the desire to preserve ownership of inherited land, sheltering nonfarm income from taxation, or holding land in anticipation of development. Only from the narrow, shortrun viewpoint of current rate of return on investment can these farms be said to be performing below norms. Because of these more diverse goals and competing demands on the operator's time, these smaller farms frequently utilize their resources much less intensively than larger farms and have much lower crop yields and livestock performance rates. To the extent that agricultural resources under the ownership and control of these small units remains stable, their productivity is likely to remain low, their enterprise mix to reflect less traditional types of production, and their contributions to agricultural output to be minor relative to their resources. Their primary characteristic, however, will remain their reliance on off-farm income sources. The demographic, economic, and employment characteristics of these small farm families are nearly indistinguishable from the general nonfarm population. Their financial strength and creditworthiness are based on their high equity in farm assets and their nonfarm sources of income.

For the lower-medium size and larger commercial farms, farm profitability, growth, and consolidation are more prevalent as goals. Productivity, profitability, and intensive use of farm resources characterize the goals and operational strategies of most of these farm families. For these farms the life-cycle still tends to follow the pattern of (1) beginning on rented land, (2) acquiring ownership while continuing to farm rented land during expansion, (3) relinquishing rented land as the farm family passes in to a new generation, and finally (4) renting out or selling the farm to a new generation of operator family (Gale 1994, 1996). If farms are consolidated into larger operations between generations, their resources will be used much more intensively and productively, leading to further increases in productivity and expansion of supply. Solvency and profitability of the farm business, exposure to risks, and, to a lesser extent, supplemental sources of income are important lender criteria for evaluating the creditworthiness of these farms.

Financial Capital. The flexibility of farm firms to adjust to volatile prices, yields, and incomes is greatly reduced if all factors of production must be compensated each year. Hired labor costs must be met each year, but operator and family labor returns need not. Rental contracts must be met each year, but returns to owned land do not. Leased machinery and equipment requires annual contractual payments, but machinery depreciation does not. Financial capital considerations arise because debt capital requires a contractual payment each year, while equity capital, receiving a residual return after other expenses are covered, does not.

Nevertheless, questions of ownership versus rental of assets, equity versus debt capitalization, and growth versus consolidation of the firm do not have cut-and-dried answers. Most combinations are possible and many are feasible. Feasibility of financial organizations of farms of all sizes can generally be reduced to two considerations: cash flow, considering all sources of farm family income, and solvency, considering all family assets. In devising the cash-flow and solvency classification for farm firms in the 1980s, ERS researchers found that net cash flows tended to become very low to negative as debt/asset ratios increased beyond 40 percent. The inclusion of off-farm income sources and nonfarm assets might alter the cutoff points because the ERS classification only considers farm income sources and farm asset/debt structure. However, the 40-percent debt/asset ratio still appears to be a good cutoff point for delimiting desirable solvency of farm businesses—only 7.6 percent of farms in 1993 had debt/asset ratios higher, and over half of those also had negative cash flows.

Industry Evolution in Agriculture

Agricultural Markets Grow Slowly. In general, agriculture is a mature industry that can expect slow to moderate growth for its products. Industries anticipating rapid growth on the upside of the product cycle generate demands for large infusions of capital investment. For example, tight markets for computer chips in the early 1990s led to huge investments in new plants by chip makers in an effort to expand capacity. In contrast, markets for most agricultural products are established and grow slowly. Some scenarios, usually involving growing populations and changes in tastes (for example, increased meat consumption in China), suggest dramatic growth in food demand that will outstrip supply; but even more dire predictions of world food shortages in the 1970s failed to materialize, largely because of the Green Revolution. Instead, the result was excess capacity during the 1980's. The boom and bust cycle of the 1970s and 1980s illustrates the problems that can occur when economic actors misinterpret transitory market disruptions (e.g., Russian grain sales in the 1970s) as permanent changes in market conditions that justify capital investment. Boom-bust cycles are not unique to farming. There is evidence of a similar boom-bust cycle in the computer chip industry in the mid-1990's. Chip prices plunged as new plants came into production, and some of the newly constructed plants are idle due to excess capacity.

Long-term aggregate trends in U.S. and global food production and demand have proven reasonably predictable. As previously mentioned, growth of effective demand and improvement in agricultural productivity have to be balanced in order to maintain price stability. In most postwar periods, growth in production capacity has outstripped growth in effective demand. Occasional crises—such as the Russian entry into the grains market in the early 1970s, the Sahel drought in the 1970s, and demand spurts in certain developing countries caused by the influx of Eurodollar investments—were all traceable to specific, transitory causes. The demonstrated ability of U.S. and developed world agriculture to expand production faster than demand has always reasserted itself. In the short run, local or more widespread disruptions in weather and markets, such as El Niño, can cause severe volatility of prices.

Production Adjusts Continually. Slow growth means that the world market for agricultural products is characterized by competition among producers for shares of a relatively static market. The competition occurs through government interventions (various subsidies, trade and macro policies) and through producer adjustments that increase efficiency. As trade liberalization in agriculture takes effect, producer adjustments may become more important. U.S. commercial-size farms, emerging from difficult times in the 1980s, have pared production costs by adopting new technologies, substituting machinery for labor, and operating more acres or animal units per farm. The increases in size of commercial farms reflect the desire to spread labor and machinery costs over more units of output. Increasing farm size in a slowly growing market for farm products and a fixed agricultural land base necessarily results in a declining number of farms and farmers. The total commitment of financial resources to agriculture has not declined, however, and will continue growing, despite the continuing decline in farm numbers. Among larger farms, operators will continue to face competitive pressures that will induce them to adopt new advances in machinery, technologies, and management practices and further spread capital investment over more units of output. This will lead to further increases in capital investment per farm —whether through ownership, leasing, or contractual relationships.

But, Much of Structural Adjustment Is Intergenerational. Much of the adjustment to fewer, larger farms has occurred gradually through generational change. The many small, diversified farms of the 1950s generation have been replaced by fewer, larger, and more specialized farm operations in the 1990s. Intergenerational farm succession in the United States contrasts with the problem faced in less developed economies, where a family's farmland may be divided among the offspring upon the family head's retirement. In the United States, relatively few young persons are entering farming to replace retiring operators. The land of several retiring operators may be consolidated by purchase or lease into a single operation controlled by a younger farmer. Much of the increase in farm size can be accounted for by larger entering farms replacing smaller exiting farms (Gale 1994, Gale and Pursey 1995).

The Evolving Policy Environment

Agriculture is losing some of its uniqueness. As agricultural production becomes more concentrated among fewer and larger farms, it takes on more characteristics of other commercial sectors. As the bulk of small commercial and noncommercial farms rely more heavily on off-farm income sources, this portion of the agricultural population increasingly resembles the general population. The 1996 Act may be seen as a turning point in policy toward the agricultural sector. It may signify that the Congress no longer sees either commercial agriculture or small farm agriculture as needing their former levels and types of assistance and support.

Macro and Trade Policies. Governmental policies toward the agricultural sector are currently shifting toward harmonization with policies toward other commercial sectors. Macro and trade policies are becoming more important to the economic well-being of farms than traditional commodity programs. The farm financial crisis of the 1980s was largely caused by an abrupt change in macro policy that, first, greatly increased real interest rates, thus increasing operating costs and reducing asset values, and second, increased the value of the dollar in international exchange, thus restricting U.S. export markets and stimulating imports of competing foreign goods. These changes in macro policy made unsustainable the trajectories of farm incomes and asset values that had built up over the 1970s (Duncan and Harrington 1986). Commodity programs were not well targeted to deal with the resulting farm financial stress; and the financial stress spread from commodity sector to commodity sector, to the farm service sectors, machinery manufacturers, lenders, and finally rural communities. In spite of record-level government transfers to the farm sector, the deepening of the financial stress could not be slowed nor the recovery accelerated, demonstrating the inadequacy of commodity programs to deal with macro policy problems.

The inclusion of agriculture in the General Agreements on Trade and Tariffs (GATT, superceded by WTO) negotiations for the first time in the recently completed Uruguay Round and the recent negotiation of NAFTA further demonstrate that agriculture, worldwide, is no longer considered unique in trade matters. The United States entered into these agreements anticipating improved access to international markets and increased effective demand for U.S. agricultural products. Indeed, significant expansion of agricultural markets is necessary to achieve the goals of the 1996 Act without major economic dislocations in agriculture. Recent trade growth shows that the envisioned expansion is likely. World trade in agricultural commodities and processed foods increased nearly sixfold between 1972 and 1993, an annual growth rate of almost 8 percent (Henderson, Handy, and Neff 1997, pp. 7-8). Processed foods now account for two-thirds of total agricultural trade and represent the largest industrial sectors of most national economies (13.5 percent of U.S. manufacturing output, and from 33.7 percent for New Zealand to 9.8 percent for Japan). The food trade in agricultural commodities and processed products is big business, nearly indistinguishable from other industrial and commercial sectors.

Tax Policies. Two policy areas where agriculture retains some of its uniqueness are in tax and succession policies. Farming has traditionally been, and continues to be, a tax-favored activity. In federal income taxation, most farmers are allowed to use cash accounting methods, which allow them to prepay expenses in order to delay or avoid the recognition of income for taxation (Davenport, Boehlje, and Martin 1982; Freshwater and Reimer 1995). Farmers may also convert current income into unrecognized capital gains through direct expensing of some equipment investments, and by writing off as business expenses interest paid on investments in appreciating assets, such as land. The implications of these income tax provisions are that farms have strong incentives to continuously increase their investment in order to further delay the recognition of income, giving a strong bias toward farm growth and increasing financial risks in unfavorable income years. The deferred income taxes come due if the farm's income is too low to continue prepaying expenses in any year. And, the deferred capital gains taxes come due if the owner is forced to sell farmland in order to weather a period of low income.

The federal estate tax also has provisions that treat farms uniquely. One of them is the "stepped-up basis"—the valuing of inherited property at the value of the assets at the time they are inherited rather than acquired. Because of this, capital gains on assets occurring during the lifetime of a farm owner, much of which may have been paid for with tax-deductible interest expenses, are never recognized and taxed. Estate taxes are due only on inheritances above a generous exemption, increased to $1.2 million in 1997. Further, qualified heirs may calculate the value of the estate at the agricultural value of the assets—usually considerably less than their fair market value, especially in urban fringe areas. Finally, qualified heirs can delay estate taxes for five years interest-free, and pay them over an additional ten years at preferential interest rates.

The implications of these tax treatments, unique to agriculture, are:

- More resources are attracted into or retained in farming than would otherwise be the case, resulting in depressed market returns and higher costs and values of assets that can benefit from the tax sheltering provisions.
- More farm assets are held until death to avoid capital gains taxes and take advantage of the stepped-up basis. This leads to more rental of agricultural assets, especially share rental arrangements that preserve an estate's qualification for the preferential valuation and deferred payment of estate taxes.
- The age structure of farmers and the number of small farmers and retired farmers are biased upward. In transferring their farm to the next generation, many operators may consider themselves retired, rent out most of their assets, but still retain control of enough assets to qualify as noncommercial farms.
- Current net incomes and rates of return in farming are depressed below what they otherwise would be, while capital values of agricultural assets are biased upward. This tendency is pervasive enough to lead to the well-known aphorism, "Farmers live poor and die rich."

Environmental Policies and Resource Programs. Another area in which agriculture retains some uniqueness is in environmental and resource policies. Agriculture remains the only sector of the economy in which a significant share of environmental and conservation expenditures are underwritten by federal and state assistance. In other industrial sectors, most of the burdens of environmental measures are borne by the industries (or shifted to consumers). The 1996 Act continues this tradition of underwriting a large share of agricultural environmental measures in the Conservation Reserve Program (CRP), as well as the education, technical assistance, and cost-sharing provisions under the Environmental Quality Incentives Program (EQIP) . This unique treatment may be because agriculture is characterized by large numbers of farms spread over most of rural America, and is not as amenable to regulation and monitoring as other, more geographically concentrated sectors.

Resource and environmental programs under the 1996 Act continue to fund the CRP at $1.9 billion per year for up to 36.4 million acres. The Act also provides for consolidation and simplification of environmental and conservation programs under EQIP, funded at a total of $1.3 billion over seven years. Both the CRP and EQIP funds are intended to obtain the maximum environmental benefits per dollar spent.

Resource and environmental regulations—including zoning and development rights, land use, pollution abatement, and endangered species regulations—are of paramount importance to states and the nation as well as farmers and lenders. The implications of environmental and resource regulation policies for property rights of land owners, cost of operation of farms, and liabilities for damages are complex and vary from locality to locality because of state jurisdictions and regulations, as well as local resource endowments (e.g., coastal zones, wetlands, fragile lands, and endangered species habitats). Even if federal commodity policies and programs are suspended entirely, federal and state resource and environmental policies and assistance programs will continue to be major forces shaping the agricultural sector and major factors to be considered in agricultural finance.

Commodity Programs. With the passage of the 1996 Act, Congress clearly indicated that agricultural commodity programs were in transition; but the question remains, "In transition to what?" The production flexibility contract (PFC) payments are currently scheduled to end with the expiration of the 1996 Act in 2002. Whether or not these payments end may depend on how the agricultural sector uses its temporarily increased transfer payments, how it adjusts its greater production flexibility to the expected growth in markets , and how it weathers any resulting economic dislocations. The authority for continuing commodity programs still exists, inasmuch as the "permanent legislation" is only suspended for the duration of the 1996 Act. Congressional budget projections still include a minimum of $4 billion in annual farm program costs after the expiration of the 1996 Act in 2002. Nevertheless, if a majority of the Congress feels that the transition payments were meant to 'buy out" agriculture's claim to continuing governmental support and assistance, it will be very difficult to reinstitute commodity programs in a form resembling previous programs.

Risk Management Programs, Public and Private. The volatility of agricultural prices and the resulting risk for producers and lenders may remain a problem, even with new revenue insurance and futures/options market tools envisioned in the 1996 Act. Futures and options markets are designed to transfer risks arising from short-term variability of prices. But, they entail some risks of their own if not used skillfully. Futures and options trading alternatives are complex and varied, providing risk shifting opportunities that range from:

- Fixing futures prices, bases, and delivery quantities (forward selling, hedging a growing crop),
- Fixing only the basis (basis contracts),
- Fixing only the futures price (hedge-to-arrive contracts),
- Acquiring only the option (but not the obligation) to sell or buy a quantity of commodity (put or call options).

Each of these alternatives entails a different level and type of risk for the producer, and some combinations increase rather than decrease the producer's revenue risk or the buyer's cash flow requirements—as demonstrated by the losses sustained in hedge-to-arrive contracts in 1995-1996 (Wright and others 1996). Nevertheless, when combined with crop yield insurance or used as the pricing mechanism in revenue insurance, futures/options activities become very effective in reducing the interyear variability of crop revenue (Harwood and others 1997).

Medium-term (up to five-year) variability of crop revenue was the focus of the original proposals for revenue insurance. Revenue insurance as proposed in the "Iowa Plan" used moving average prices and moving average yield histories as the bases to compute crop revenue for insurance purposes. The Iowa Plan focused on the short term variability of prices by using a five-year price and yield averaging period. Other proposals for revenue insurance, such as the Harrington-Doering proposal (Harrington and Doering 1993) and the Canadian GRIP programs used longer price and yield averaging periods to focus on longer term revenue risks.

Long-term trends and variability in revenue, such as occur with differential growth rates of effective demand and productive capacity, cannot be addressed by any of the yield, price, or revenue insurance tools. Entry/exit of farms and resources from farming, shifts among production specializations, and induced technological change are still the only responses that deal with long-term secular trends in agricultural profitability.

The Emerging Future

The future course of the agricultural sector depends on balancing growth of markets and growth of agricultural production capacity, which in turn influences the level and composition of income, asset values, and farm numbers. ERS research and projections made after the passage of the 1996 Act are used in this view of the future. These projections compare the expected performance of the agricultural economy to 2005 under the 1996 Act with what it would likely have been under previous legislation—continuing past trends, linkages, and relationships in resource use, production, trade (including implementation of the GATT and NAFTA trade agreements), and consumption — assuming no shocks due to weather, macroeconomic developments, or further international events.

Effects of the 1996 Act are mixed for many commodities and the overall impacts of the Act are projected to be relatively small. Crop production patterns are not projected to change greatly and export market competitiveness of U.S. production is projected to remain high, with the dollar value of U.S. agricultural exports rising by nearly one-third to $80 billion by 2005.

Income Levels and Composition. With the projected robust export growth and limited response of U. S. production, net farm income (Figure 2.8) initially falls from the near-record levels of 1996 to the trend of the recent past, then stabilizes around $42 billion (Young and Westcott 1996). After 1997, income levels under both the continuation of previous legislation and the 1996 Act rise slightly, with income under the 1996 Act initially $4 billion higher than under previous legislation in 1996, then $2 billion higher in later years. This reflects the increases of PFC payments over what would have been paid as deficiency payments. With the growth in export markets, there is little change in net farm income if the Act expires unextended in 2002.

Even though the effects of the 1996 Act are projected to be relatively minor at the national level, the transition to lower or eliminated direct government payments will have different effects by region, commodity, and dependence on farm revenue sources. The farms that will experience the most adjustment pressures are those that currently depend heavily on government payments. In 1995, there were about 34,000 farms that depended on direct government payments for over 20 percent of their gross farm income (FCRS data). These included wheat/cotton farms in the Southern Plains, wheat farms in the Northern Great Plains, and

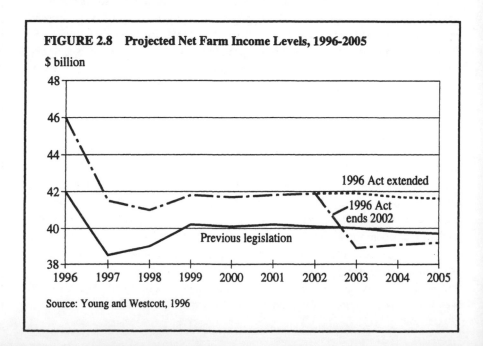

FIGURE 2.8   Projected Net Farm Income Levels, 1996-2005

Source: Young and Westcott, 1996

mixed grain farms in the western Corn Belt. These farms tend also to be in areas where access to off-farm employment is not as strong as in the rest of the country, which will exacerbate their adjustment pressures. In addition, with the phase-out of dairy price supports, dairy farms in the North Central and Northeast regions may experience adjustment pressures. Their opportunities for off-farm employment may facilitate their exit from dairying, while remaining in farming in another specialization.

The composition of farm family incomes will continue to reflect strong reliance on off-farm income sources for the bulk of farm families operating non-commercial and small commercial farms. The vitality of the local economy is the most important factor in the composition of their incomes. Though local economic vitalities are influenced by macroeconomic policies, they are not directly affected by the 1996 Act and are not projected to change from current conditions. Regional variations in off-farm employment opportunities are likely to be the most important factors determining the severity of the impacts if the 1996 Act expires unextended.

Land Values. The projections of land values have been made under two alternative assumptions because of their dependence on producers' expectations for continuation of the programs, as well as the current levels of net farm income. The 1996 Act will initially raise land values above what they would have been under previous legislation, perhaps as much as 5 percent higher by 1998 (Figure 2.9) (Young and Westcott 1996). Land values vary widely from this national average based on both farm income expectations and nonfarm opportunities such

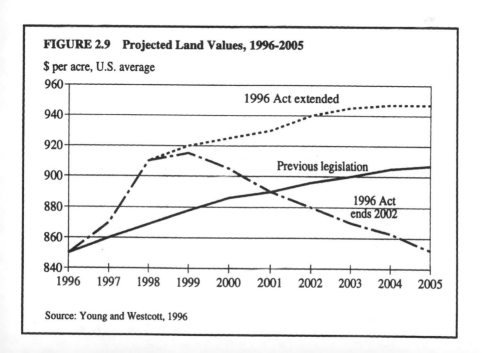

FIGURE 2.9 Projected Land Values, 1996-2005

$ per acre, U.S. average

Source: Young and Westcott, 1996

as urban pressures and development opportunities. The stronger the urban and development pressures in a local area, the less land values will be affected by the changes entailed in the 1996 Act. After 1998, if the Act is continued, land values will stay about 5 percent above what they would have been. If, on the other hand, the PFC payments end in 2002, farmers will start to anticipate this outcome and reduce the amounts they are willing to bid for land. This will cause land values to erode starting in 1999, drop below those of the previous legislation around 2001, and finally track about 5 percent below those of the previous legislation by 2004.

Farm Numbers. The number of U.S. farms is expected to continue declining over the coming decade. Of the 1.9 million farmers counted in the 1992 Census of Agriculture, over 900,000 were at least 55 years old. Analysis of historical data shows that about half of farmers in that age cohort will leave farming over a ten-year period (Gale 1996). Between 1992 and 2002, members of this oldest cohort are expected to decline by 533,000, and middle-aged farmers who were age 40-54 in 1992 will decline an additional 78,000. The youngest cohorts will add about 160,000 new farmers. Gale (1996) projects a decline to 1.48 million farms in 2002, and a further decline to 1.29 million in 2007 (Figure 2.10). Pressures to consolidate farm operations, as described in previous sections, produce the imbalance between young entrants and older retiring farmers. Entry barriers, including high capital requirements and lack of credit, may keep out some beginning farmers, but the low rates of farm entry are mostly due to poor earn-

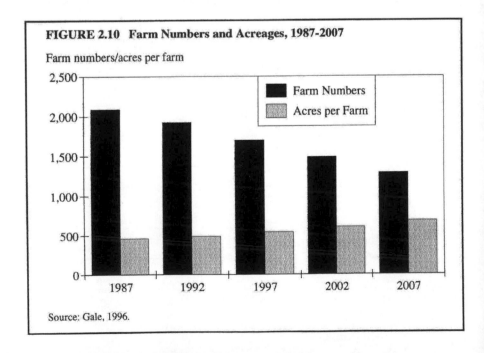

FIGURE 2.10 Farm Numbers and Acreages, 1987-2007

Farm numbers/acres per farm

Source: Gale, 1996.

ings prospects in farming compared with other occupations. Again, these trends are largely based on commercial farms. Strong entry of noncommercial farms accompanying the "rural rebound" of the 1990s may offset the expected decline in commercial farms. Unusual gains in farm prices and incomes could also offset this trend by inducing more entry, as occurred in the 1970s.

One of the key issues is how to accomplish the intergenerational transfer of farm assets in an equitable and efficient manner. Farmers age 55 and older in 1992 controlled about 460 million acres of farmland, and the value of farmland and buildings controlled by this cohort was $318 billion (Gale 1996; 1992 Census of Agriculture). Partly because of tax incentives mentioned earlier, and partly because of the farm financial crisis of the 1980s, control of farmland has become more concentrated in the hands of older farmers. Between 1982 and 1992, the land in farms controlled by farm operators age 55 and older rose 51 million acres, and the share of land controlled by that age group rose from 44 to 49 percent. The share of land controlled by farmers age 65 and over increased from 17 to 24 percent over the same period. These trends are inducing state and federal governments to increase credit availability for beginning farmers and offer innovations such as the "land-link" programs initiated in a number of states that match potential farm entrants with retiring farmers who have no heir. High capital requirements also contribute to the popularity of contracting relationships between farmers and agribusinesses.

However, from a total production perspective, many of the operators over age 65 do not have to be replaced as they leave farming. They have already left farming to a large degree. They classify themselves as retired, but are counted as farms because they continue to operate assets capable of producing $1,000 per year in sales of agricultural products. The replacement of the production of these operators by younger operators has mostly already occurred. If they are not replaced, only the 2 percent of production they account for would be lost to the sector.

Summary Implications for Agricultural Finance

Both long-term trends and recent trade, commodity, environmental, and resource policy changes will affect agricultural finance into the next century. A thumbnail list of the agricultural finance implications of changes in the sector and in policies follows.

• Farming will continue to evolve toward a more dualistic structure, with larger sizes of commercial farms accounting for the bulk of production, but with noncommercial and small commercial farms dominating the number of farms. Control of agricultural resources will be evenly split among sales classes because the intensity of resource use increases with farm size—with the largest farms producing over thirty times as much value of output per dollar of land and buildings than the smallest farms.

- Agricultural prices and production are likely to become more volatile as governmental price stabilization and supply management activities are curtailed. Governmental revenue stabilization activities (deficiency payments) have been replaced by (possibly temporary) direct payments unrelated to current commodity production or prices. The safety net under commodity prices (loan rates) will erode to a significantly lower level than in the past.
- Producers and their lenders will be exposed to significantly higher levels of price, yield and revenue risks. Risk management activities of producers will become extremely important considerations in lending decisions. Formal risk management planning from a wide array of possible tools—increasingly supplied by the private sector—will likely become as common as production and market planning.
- Farm investment decisionmaking will become more crucially dependent on timing. With more volatile income prospects and less governmental support, agricultural land and asset values will be more market-driven and thus more subject to swings in agricultural prices.
- Most farms, especially those in the smaller size categories, will remain highly dependent on off- farm income sources. Small to lower medium-sized commercial farms may find that permanent off-farm employment is a feasible and desirable adjustment for them as well. The economic well-being of most farm families will be more dependent on the vitality of their local communities and economies than on agricultural markets or government programs.

Notes

1. Any of the entities at the business level, except nonfarm contractors, may receive government payments.

2. Milk producers generally have verbal agreements with their buyers or cooperatives, but since neither the quantity nor the final price is specified, many survey respondents did not consider these arrangements as contracts.

3. The ERS financial solvency and profitability categories are based on the combined net income and debt/asset ratios of farms. Farms with positive net incomes and debt/asset ratios of no more than 40 percent are termed "favorable"; those with negative net farm income, but debt/asset ratios not more than 40 percent are termed "marginal income"; those with positive net income, but debt/asset ratios above 40 percent are "marginal solvency"; and those with both negative net farm income and debt/asset ratios above 40 percent are "vulnerable".

4. Noncommercial = sales of less than $50,000, Small Commercial = sales of $50,000 to $99,999, Lower Medium Commercial = sales of $100,00 to $249,999, Upper Medium Commercial = sales of $250,000 to $499,999, Large Commercial = sales of $500,000 to $999,999, and Superlarge Commercial = sales of over $1,000,000.

References

Ahearn, Mary C. 1986. *Financial Well-Being of Farm Operators and Their Households.* AER-563. U.S. Dept. Agr., Econ. Res. Serv., Sept.

Ahearn, Mary C. 1992. "Historical Perspectives on Measuring the Incomes of U.S. Farmers". Unpublished paper U.S. Dept. Agr., Econ. Res. Serv.

Ahearn, Mary C., Jet Yee, Eldon Ball, and Rich Nehring. 1998. *Agricultural Productivity in the United States.* AIB-xxx. U.S. Dept. Agr., Econ. Res. Serv.

Cochrane, Willard W. 1993. *The Development of American Agriculture: A Historical Analysis.* Minneapolis: University of Minnesota Press.

Davenport, Charles, Michael Boehlje, and David B. H. Martin. 1982. *The Effects of Tax Policy on American Agriculture* AER 480, U.S. Dept. Agr, Econ. Res. Serv., Feb.

Duncan, Marvin, and David H. Harrington. 1986. "Farm Financial Stress: Extent and Causes" in *The Farm Credit Crisis: Policy Options and Consequences,* Texas Agricultural Extension Service Bulletin No. B-1532, Febr.

Freshwater, David, and Bill Reimer. 1995. "Socio-economic Policies as Causal Forces in the Structure of Agriculture" in Harrington, David H., and others, Eds. *Farms, Farm Families and Farm Communities. Special Issue Canadian Journal of Agricultural Economics,* Dec.

Gale, H. F., Jr. 1996. "Age Cohort Analysis of the 20th Century Decline in U.S. Farm Numbers." *Journal of Rural Studies* 12: 15-25..

_____. 1994. "Longitudinal Analysis of Farm Size Over the Farmer's Life Cycle." *Review of Agricultural Economics* 16: 113-123.

Gale, H. F., Jr., and Stuart Pursey. 1995. "Microdynamics of Farm Size, Growth, and Decline: A Canada-U.S. Comparison." in Harrington, David H., and others, Eds. *Farms, Farm Families and Farm Communities. Special Issue Canadian Journal of Agricultural Economics,* Dec.

Harrington, David H., and Otto C. Doering III. 1993. "Agricultural Policy Reform: A Proposal," *CHOICES,* March.

Harwood, Joy, and others. 1996 "Strategies for a New Risk Management Environment" *Agricultural Outlook,* AO-234, Oct.

Henderson, Dennis R., Charles R. Handy, and Steven A. Neff, Eds. 1996. *Globalization of the Processed Foods Market.* AER 742, U.S. Dept. Of Agr, Econ. Res. Serv., Sept.

Hoppe, Robert A., and others. 1996. *Structural and Financial Characteristics of U.S. Farms, 1993: 18th Annual Family Farm Report to Congress.* AIB-728. U.S. Dept. Agr., Econ. Res. Serv., Oct.

Morehart, Mitchell J. 1995. "Financial Performance of Farms," *Agricultural Outlook 1995.* Misc. Pubn. U.S. Dept. Agr., World Ag. Outlook Bd., Feb.

Nelson, Frederick J., and Lyle P. Schertz. 1996. *Provisions of the Federal Agriculture Improvement and Reform Act of 1996.* AER 729. U.S. Dept. Agr., Econ. Res. Serv., Sept.

Olson, Kent D., and Stanton, B.F. 1993. "Projections of Structural Change and the Future of American Agriculture", in Arne Hallam, Ed., *Size, Structure, and the Changing Face of American Agriculture.* Boulder: Westview Press.

Perry, Janet, and Mitch J. Morehart. 1994. "Characteristics of Commodity Program Recipients," *Agricultural Income and Finance Situation and Outlook Report,* AIS-55, U.S. Dept. Agr., Econ., Res. Serv., Dec., pp. 18-25.

Peterson, R. Neal. *Measuring Real Size Distributions in U.S. Agriculture.* Forthcoming publication, U.S. Dept. Agr., Econ. Res. Serv.

Reimund, Donn A., and H. F. Gale, Jr. 1996. *Structural Change In the U.S. Farm Sector 1974-1987: 13th Annual Family Farm Report to Congress.* AIB-647. U.S. Dept. Agr., Econ. Res. Serv., Oct.

Stanton, B. F. 1993. "Changes in Farm Size and Structure in American Agriculture in the Twentieth Century," in Arne Hallam, Ed., *Size, Structure, and the Changing Face of American Agriculture.* Boulder: Westview Press.

U.S. Department of Agriculture. 1996. *Farmers' Use of Marketing and Production Contracts.* AER 747. U.S. Dept. Agr., Econ. Res. Serv., Dec.

Wright, Bruce, and others. 1996. "Hedge-To-Arrive Contracts: Risk and Lessons" *Agricultural Outlook,* AO-234, Oct.

Young, C. Edwin, and Paul C. Westcott. 1996. *The 1996 U.S. Farm Act Increases Market Orientation.* AIB-726. U.S. Dept. Agr., Econ. Res. Serv., Aug.

PART TWO

FUTURE DIRECTIONS FOR AGRICULTURAL FINANCE

3

New and Changing Rules of the Game

Mark Drabenstott and Alan Barkema

A sweeping transformation of the agricultural finance landscape is triggering new pressure to rewrite old rules governing agricultural lending. New financial products are evolving to fit the changing needs of a more industrialized agriculture. Vast technological innovations in information processing and communications are boosting the quantity and quality of financial information available to both borrowers and lenders. And new competitive pressures are emerging among agriculture's traditional lenders and newer entrants to the field. Thus, the period ahead may witness a substantial rethinking and perhaps an overhaul of agricultural finance regulations.

One factor that will drive the re-examination of these regulations is the growing recognition that traditional boundaries surrounding agricultural finance are fading away. In the past, agricultural finance has typically been defined as providing debt capital to agricultural producers. Today, the lines between producer and processor are blurring, as more of the nation's food production falls under production or marketing contracts. Thus, regulations must recognize that agricultural finance increasingly involves the whole sector, and no longer just production. Moreover, as the structure of the industry undergoes rapid change, capital needs are becoming more complex. Traditionally, agricultural finance regulation has focused almost entirely on debt capital. In the future, other forms of finance will also be important.

This chapter considers three key questions emerging from these developments in the agricultural credit market. *How is the agricultural capital market changing?* The first section shows that the agricultural credit market is now growing again but remains highly competitive, and the competition is starting to bring some pressure to bear on the rules that govern farm lending. *What implications do these changes suggest for the market's regulation?* The second section shows the implications may be greatest for commercial banks, which have the heaviest net of regulations, although recent changes in Farm Credit System (FCS) rules will invite new debate on society's contingent liabilities in a single-sector government-sponsored enterprise (GSE). *What regulatory changes may be anticipated in the years ahead?* The chapter concludes that some rebalancing of costs and benefits in commercial bank regulation is likely, that the public will take a wary eye of GSE involvement in rural America, and that new ways may emerge to assess the risk of lending to agricultural production under contract to processors.

How Is the Agricultural Finance Market Changing?

Three key factors characterize the changes taking place in the agricultural credit market of the late 1990s. First, the market for agricultural credit—traditionally the primary focus of agricultural finance—is growing again, although relatively slowly. Second, competition in agricultural lending has intensified as old competitors have gained new financial muscle and new competitors have emerged in the already crowded field. And third, the growing need for equity capital points to the possible introduction of a new generation of agricultural finance products to meet that demand.

The agricultural credit market remains a slow growth market, despite an uptick in recent years. During the five years ending in 1997, debt owed by U.S. farmers and ranchers grew by about $20 billion, ending nearly a decade of decline or no growth in farm debt. At an expected $160 billion in 1997, however, farm debt still remains well below the $193-billion peak of the early 1980s. Looking ahead, the pace of farm borrowing is difficult to predict. Recent strength in farm exports may rekindle exuberant expectations of farm incomes and trigger new borrowing to support additional farm investments. But volatile commodity markets and the phasing out of the government safety net for farm income may quickly deflate such exuberance. On balance, the recent upturn in farm borrowing may be promising to farm lenders, but it is too soon to predict a renewed surge in farm borrowing.

Nevertheless, the slowly growing farm lending market is crowded with eager competitors. As a group, commercial banks are the leading farm lender. Since the end of the 1980s farm recession, they have been especially aggressive and successful competitors, nearly doubling their share of farm debt to two-fifths while boosting their profits to record or near-record levels (Figure 3.1). Agricultural loans are made by nearly all sizes of commercial banks. Indeed, the commercial bank portion of the agricultural credit market is fairly evenly divided among banks of widely varying size (Table 3.1). The smallest banks, those with assets under $50 million, control about a fifth of the total loans made to agriculture by commercial banks. Banks with $50-$100 million in assets, which are still quite small as financial institutions, control another fifth. Midsized

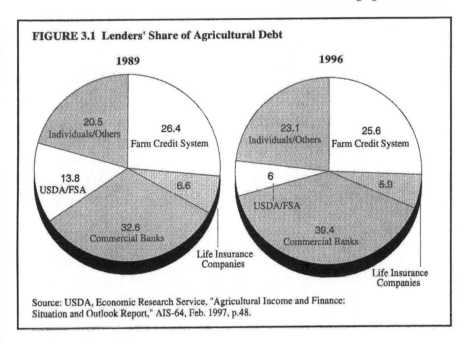

FIGURE 3.1 Lenders' Share of Agricultural Debt

Source: USDA, Economic Research Service, "Agricultural Income and Finance: Situation and Outlook Report," AIS-64, Feb. 1997, p.48.

banks, those with $100-$250 million, also control about a fifth of the commercial bank loans to agriculture. Finally, the largest banks, those with assets greater than $1 billion, control about a quarter of bank loans to agriculture. Of note, agricultural banks, those with a farm loan/total loan ratio that exceeds the nation-

TABLE 3.1 Agricultural Loans at U.S. Commercial Banks, December 31, 1996

Size of Bank (assets)	Agricultural Loans	Percent of Commercial Bank Agr. Loans	Agricultural Loans as Percent of Total Assets
	Thousand dollars	-----*Percent*-----	
< $50 Million	14,317,533	21.86	13.29
$50 - $100 Million	13,817,805	21.09	7.99
$100 - $250 Million	13,150,144	20.08	4.24
$250 - $500 Million	4,433,713	6.77	2.06
$500 Million - $1 Billion	2,901,090	4.43	1.53
> $1 Billion	16,884,505	25.78	0.57
Agricultural Banks*	36,113,040	55.13	19.64

* Agricultural banks have agricultural loans comprising a higher percentage of total assets than the unweighted average of this ratio for all commercial banks in the U.S, currently 16.3 percent.

Source: Calculated from the Report of Condition and Report of Income files, Board of Governors of the Federal Reserve System.

al average of 16.3 percent, remain a strong force in agricultural lending, controlling 55 percent of all commercial bank loans to agriculture. These banks have been quite successful in the 1990s, and have added to their market share, relative both to other banks and to the FCS.

Another important, though unquantified, part of the gain by banks has been by international banks now operating in the United States. Banks such as Credit Agricole (France) and Rabobank (Netherlands) have become active participants in the market for agricultural loans, especially bigger credits. Data to document the size of their presence is, however, unavailable at the present time. Nevertheless, anecdotal information received by the authors suggests these banks are major players in some regions of the country, tailoring their operations to larger farm and agribusiness credits that can be serviced efficiently with streamlined loan production offices.

The entrance of international lenders raises a number of interesting regulatory questions in the agricultural finance arena. Many of these international banks enjoy regulatory freedoms that U.S. banks do not. For instance, ING-Barings is a major international financial service company that now makes agricultural loans in the United States. This firm began as a Netherlands bank, but it has merged with a British investment bank (Barings) and is currently in the process of buying U.S. insurance companies. Moreover, continued growth in the United States of an institution such as Credit Agricole, a cooperative lending institution based in France, could prompt new questions about the charter of the FCS, itself a cooperative. Credit Agricole has broad authority to lend to all industries and in any country it wants, unlike the FCS, which still is a single-industry, single-country lender.

Much of the commercial banks' recent gain in market share came at the expense of the FCS, the second leading farm lender. The FCS' market share—about a fourth in 1996—is down from about a third in the early 1980s, a lasting legacy of the FCS' fall on hard times during the farm recession. Since then, however, the FCS has restructured its entire organization, rebuilt its capital base, and restored its profitability. Measured against total assets, the FCS' capital base has grown to a solid 14.5 percent and, with earnings averaging more than 1.5 percent a year, FCS compares favorably with many commercial banks (Farm Credit Administration 1997b).

Despite the marked improvement in the FCS' financial foundation during the past decade, it will face continued competitive challenges. Head-to-head competition with commercial banks and other lenders is likely to continue and perhaps intensify. In recent years, for example, FCS institutions have maintained a competitive advantage in lending to larger commercial farmers, who require credit volumes in excess of loan limits at smaller commercial banks (Economic Research Service 1997). That advantage may erode, however, as the continued restructuring of the commercial banking system creates larger institutions with the financial muscle to handle larger credits.

In addition, other lenders—including vendors, individuals, and others—are a growing presence in the farm lending market, attracted in part by new information technologies that enable low-cost lending from distant offices and by the strong earnings of the FCS and commercial banks in recent years. With new low-cost entrants crowding into the agricultural lending market, FCS operating margins could erode. Thus, competition from a growing field of low-cost, nontraditional lenders in the years ahead may be at least as important as competition from commercial banks in shaping the fortunes of the FCS.

Indeed, this miscellaneous group of nontraditional lenders is the third leading lender to agriculture, collectively holding about a fourth of farm debt. The group is widely diverse, and includes retiring farmers who provide credit to buyers of their farms and input supply firms that provide point-of-sale financing for their customers. Recent innovations among input suppliers, such as machinery dealers and farm chemical and seed companies, point to increased interest in farm lending by this group. Many of these firms now recognize lending as an integral part of their overall relationship with their customers, and they are developing their credit operations as both a potent marketing tool and a profit center.

Unfortunately, data are not available to gauge the overall size of the farm loan portfolio held by merchants. Nevertheless, available data for a few input companies suggest their market share is expanding rapidly and significantly. Pioneer Credit, for example, grew its loan portfolio from just $2.5 million in 1989 to $115 million in 1997. John Deere Credit has grown from $2.3 billion in 1987 to $3.8 billion in 1997. Such growth puts merchant credit squarely on the map as a key source of credit, and owing to the very different nature of these lenders, raises some interesting comparisons on regulatory matters.

Life insurance companies also remain a competitive force in the farm lending market, primarily in the market for larger farm mortgages. While life insurance companies hold a relatively small slice of farm debt—less than a tenth—they remain strong competitors in their market niche, which they serve efficiently with lean organizations (Stam et al. 1995).

While credit remains the main thrust of agricultural finance, the sector's capital needs are becoming more complex, posing new questions about the rules that have traditionally governed the U.S. agricultural finance market. For example, many farmers and rural communities are investing in food processing plants to add more value to bulk commodities before they leave the local community. Such plants are highly capital-intensive, and thus equity capital is needed for such efforts to be successful. Rural equity capital markets, however, are not well-developed, and most states will be looking more closely at the rules that support the development of such markets.

Overall, the outlook for the agricultural lending market suggests the presence of numerous lenders with the financial muscle to boost lending. But the overall market may continue to grow slowly, and as a result, competition for market share will remain intense. Inevitably, the intense competition brings pressure to bear on the framework of rules and regulations that govern farm lending. The emergence of international lenders and the growing complexity of agriculture's capital needs will only add to that pressure.

What Pressures Are Building for Regulatory Shifts in Farm Lending?

Financial regulations will remain an important factor in the new competitive environment in agricultural finance. Agricultural lenders are subject to regulation, but the degree of regulation varies widely. One critical distinction can be drawn between public and private charters. Publicly chartered institutions, such as commercial banks and the FCS, have their own set of regulations. These regulations include safety and soundness criteria that attach to the special authorities granted in the public charter. Privately chartered institutions are subject to a different set of regulators and regulations. These rules address some of the same concerns, but in the main are aimed at ensuring that investors and others have full information on the financial condition of the company.

Regulations for Publicly Chartered Institutions

In the public arena, regulations for agricultural lenders have typically been guided by three principal goals or objectives: to ensure the safety, soundness, and stability of the financial system; to promote financial market efficiency and an appropriate level of competition; and to protect consumers in their individual relationships with financial institutions. Overall, financial regulations are designed to correct market imperfections and provide consumers a safer, more efficient, and fairer financial system (Federal Financial Institutions Examination Council 1992, Spong 1990).

These benefits of regulation, however, are obtained at substantial cost. Some regulations bar financial institutions from activities they might otherwise pursue, constraining innovation in the financial system and creating opportunity costs. Other regulations directly boost operating costs in financial institutions by increasing reporting requirements, paperwork, and the human effort required to ensure regulatory compliance. In the end, consumers pay for the burden regulations place on financial institutions through higher costs for financial services. Thus, a balance must be struck between regulatory benefits and regulatory burden (Ahrendsen and others 1995).

Commercial Banks. Among agricultural lenders, commercial banks face the heaviest set of regulations aimed at all three regulatory objectives. Safety and soundness rules have the longest history, many stemming from the financial panic of the 1930s. Among the most important safety and soundness regulations are those that restrict banking activities to avoid conflicts of interest, limit banking activities to areas of expertise, and reduce risk. Other safety and soundness rules, such as limits on loans to individual borrowers and minimum capital requirements, ensure a sound financial backstop behind consumer deposits. To further ensure safe and sound operations, banks are also required to report insider loans, balance sheets, and income statements; and to submit to periodic examination. Federal Deposit Insurance stands as a buffer between banks' business risks and consumer deposits, and the Federal Reserve System provides an ultimate source of liquidity as the banking system's lender of last resort.

As the banking system stabilized in the decades after the 1930s, regulatory concerns gradually shifted from emphasizing safety and soundness to promoting an appropriate level of competition and efficiency. Competition and efficiency regulations generally fall in two areas: chartering requirements and geographic restrictions. Chartering requirements are designed to exclude unscrupulous or inexperienced lenders. Geographic restrictions are designed to prevent excessive competition, limit concentration of banking assets, and prevent the siphoning of deposits by distant institutions with little concern for the needs of local communities. In the 1980s, however, advances in communications and electronic banking technology, the need for out-of-state recapitalization of weak institutions, and growing competition from nonbank providers of financial services led most states to relax their geographic restrictions. Recent federal legislation, which phased in full interstate branching by mid-1997, brushes aside most remaining geographic barriers.

The third regulatory objective—consumer protection—developed during the past 25 years with the growing complexity of financial products. Consumer protection rules target disclosure and fairness. Disclosure rules require banks to inform consumers of credit terms accurately and uniformly. Fairness rules ensure that credit decisions are not based on factors unrelated to creditworthiness, that banks serve the needs of their local communities, and that consumers generally are treated fairly in their relationships with banks.

As international banks continue to grow as a presence in agricultural lending, more questions may arise concerning the rules under which these banks operate. One concern is whether international banks have different capital standards. Recent legislation appears to allay such a concern. The Foreign Bank Supervision Enhancement Act of 1991 now puts foreign banks under essentially the same ground rules as domestic banks. Moreover, the Basle agreements (reached in the 1980s) to put banks throughout the world on the same capital standards largely eliminated concerns that banks in one country could become more leveraged than in another. A major contribution of the Basle accord was to make set capital standards on the basis of risk-adjusted assets. That is, assets with a higher risk of default should be backed by a higher level of bank capital. In the United States, however, the definition of "well-capitalized" remains 10 percent of assets, compared with the international standard of 8 percent. In practice, this appears to have little implication, since U.S. banks are considered "adequately capitalized" at the 8-percent level.

Another area of regulatory concern in the future will be whether U.S. and international banks are operating on a level playing field in terms of the range of activities permitted by regulations in their home country. Since the Depression era, banking legislation in the United States (the Glass-Steagall Acts of 1932-1993—P.L. 72-44 and P.L. 73-66) has prevented commercial banks from engaging in investment banking activities. Banks in Europe and Japan are not subject to this prohibition. As global capital markets have become more integrated, foreign banks have been free to enter the U.S. market with a broader range of financial services at their disposal. The question is, does this give foreign banks an inherent advantage in serving the capital needs of U.S. agriculture (or any other sector, for that matter)?

While the answer to this question is unclear at present, regulators are giving more thought to whether there are alternative regulatory approaches that preserve financial market stability while also leveling the competitive playing field. One approach that has been suggested is to require those institutions that engage in an expanding array of complex financial activities to give up direct access to deposit insurance and the government safety net in exchange for reduced regulation and oversight (Hoenig 1996). In principle, banks would give up deposit insurance in return for greater flexibility in providing financial services to their customers. Financial disclosure to investors and financial market participants—and their collective response to such information—would take the place of certain oversight activities currently undertaken by the regulators. Overall, it seems likely that the differences between bank regulation in the United States and in other countries will drive a continuing debate on how to level the playing field and maintain financial stability in global capital markets.

Farm Credit System. Regulation of the FCS targets the same three objectives as bank regulations—safety, efficiency, and fairness—with important differences arising from the FCS' cooperative structure and historical mandate as an agricultural lender. Regulations to ensure the safety of FCS institutions generally parallel bank regulations. But FCS regulations focus on the risk of default on FCS securities rather than the risk of loss to depositors. Like commercial banks, FCS institutions are required to maintain capital above minimum levels, and limits are imposed on lending to single borrowers. Other restrictions limit the FCS investment portfolio to a well-diversified group of relatively safe securities. Finally, Congress established an insurance fund as an additional line of defense between the FCS' business risks and its bond holders.

Regulation of competition is somewhat more restrictive for FCS institutions than commercial banks due to FCS' original charter as a GSE. Under that charter, FCS lending was restricted to farming and rural housing because, it was thought, this was the sector of the economy that needed new access to national capital markets. Obviously, this represents a much narrower line of business than commercial banks may pursue. Several recent proposals, however, would broaden FCS lending authority to include agribusiness, expanded rural housing, and rural development.

FCS institutions also face stringent consumer protection rules, many of which were developed in the wake of the 1980s farm recession. Like banks, FCS institutions are required to accurately disclose credit terms and allow borrowers to inspect credit documents. Moreover, FCS borrowers with problem loans enjoy substantial regulatory protection. The FCS is required to consider restructuring nonaccrual loans before foreclosing and to weigh the expected costs of alternative actions before proceeding. Borrowers are also guaranteed the right of first refusal before foreclosed property is offered for sale. Other rules similar in purpose to the community reinvestment mandate of commercial banks require FCS institutions to adequately serve the credit needs of all segments of production agriculture.

More recently, the FCS undertook some significant changes in regulations that could set the stage for ongoing debate on regulatory issues. In March 1997, the Farm Credit Administration (FCA) changed its customer eligibility rules. The changes essentially broadened authority to finance full- and part-time farmers, farm-related businesses, agricultural processing cooperatives, and rural housing. All of the changes were made as a result of FCA's re-examination of FCS's existing statute. The new rules are currently being challenged in court by the American Bankers Association and the Independent Bankers Association of America. In effect, the new rules provide new opportunities for the FCS to lend to farmers and related businesses, but may raise new questions about the public's contingent liability of a GSE still concentrated on a single sector of the economy.

The changes in farm-related businesses and agricultural processing firms could yield some additional growth in the FCS portfolio. For example, FCS loans to agricultural cooperatives grew from $6.8 billion in 1996 to $10.0 billion in 1996 (Figure 3.2). That growth came through traditional FCS channels—CoBank and the other two Banks for Cooperatives (one of which recently merged with CoBank). Under the new FCA guidelines, more parts of the FCS would be able to make loans to agricultural cooperatives. To some extent, these new FCS players will simply compete with CoBank and the St. Paul Bank for Cooperatives, which have already been making such loans. It is possible, however, that total loans by the FCS will increase. That increase will depend on the extent to which some markets were previously not served by the FCS.

Regulations for Privately Chartered Companies

Many privately chartered firms now lend to agriculture and provide other financial services. These firms stand apart from commercial banks and the FCS in that they neither accept deposits nor are they GSEs. As a result, it is not sur-

FIGURE 3.2 Farm Credit System Loan Volume Outstanding to Farm and Rural Cooperatives, 1990-1996

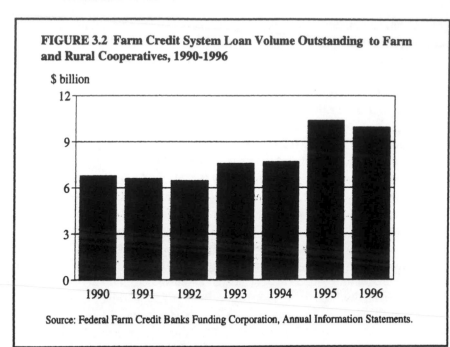

Source: Federal Farm Credit Banks Funding Corporation, Annual Information Statements.

prising that financial regulations are quite different. Two distinctions can be drawn among the privately chartered lenders.

First, vendors and merchants (such as John Deere and Pioneer HiBred) are subject to relatively little regulation. Safety and soundness of vendor firms is generally viewed as the concern of the firms' owners or shareholders, rather than depositors or taxpayers. Issuance of commercial paper and other securities by vendor firms to fund their lending activities is subject to regulatory oversight by the Securities Exchange Commission. Such a course of financial disclosure to investors essentially means that the safety of these firms is generally left to market forces. These lenders are subject, however, to credit disclosure rules similar to those that apply to commercial banks.

Second, life insurance companies face a more stringent set of rules designed to protect their policyholders' investments, which fund the companies' loan portfolios with their premiums. Life insurance companies are regulated by state insurance commissioners, who monitor company policies and impose risk-based capital standards similar to those set for commercial banks. Like vendors and commercial banks, life insurance companies are also subject to credit disclosure rules. Unlike commercial banks and the FCS, individual loans made by life insurance companies are not subject to examination and classification by regulatory examiners.

The unique features of vendor and life insurance company regulation may receive more attention in the period ahead. Vendors, in particular, pay much less to comply with regulations than either commercial banks or the FCS. Thus, these traditional lenders are likely to voice concern about whether the playing field is level.

What Changes May Be Made in Agricultural Finance Regulations?

Overall, it appears unlikely that agricultural finance regulations will undergo major changes in the next few years. If history is any guide, however, changes are certain to come when the industry enters its next downturn. The 1930s and the 1980s both witnessed substantial rewriting of agricultural finance regulations in response to a crisis in farm lending. No one can say when agricultural will enter its next downturn, but it would be highly unlikely if agriculture had suspended the cyclical pattern that has described it throughout the twentieth century. Nor can anyone say if any future downturn would be steep enough to trigger widespread concern among regulators so as to revisit the industry's underwriting rules. What does seem certain is that the players that now define the industry will spark an ongoing debate over what those rules should be. What issues seem most likely to spark debate?

Rebalancing The Costs and Benefits of Bank Regulation

There appears to be a growing recognition that rural community banks bear a bigger burden in complying with the current slate of bank regulations. One indication is the recent enactment of new rules that exempt small community banks from the full range of regulations in the Community Reinvestment Act. These changes were the result of extensive discussion on the true costs of compliance and general agreement that smaller banks bore disproportionately greater costs than large banks.

Community banks that lend to agriculture probably will continue to argue for regulatory relief in an industry undergoing substantial consolidation. As banking organizations purchase rural banks, some economists believe that lending to farmers and small businesses declines (Keeton 1996). Others believe that the consolidation has at most a neutral effect on small business borrowers (Berger and others 1997; Walraven 1997). In either case, it appears that community banks will play an even bigger role as a source of credit to these small borrowers. Recognizing that critical role, some relief in regulatory burden may be appropriate. Gilbert (1997) suggests, for instance, that relatively small banks with good supervisory ratings might be examined less frequently, thus cutting compliance costs and enabling the bank to better meet the needs of customers.

GSEs and Contingent Public Liability

Throughout the twentieth century, there have been recurring debates over the role of GSEs and the contingent liability they pose for taxpayers. These debates have been especially pitched over single-sector GSEs, such as the FCS, where cycles in the sector periodically remind taxpayers of their contingent liability. The recent widening of the FCS customer base through new FCA regulations could lead to more agribusiness loans to complement the FCS' farm loan portfolio. Still, the fact remains that the FCS remains a single-sector GSE.

Furthermore, emerging questions over the role of GSEs in meeting the capital needs of rural America will further stoke the debate on the FCS. Many community banks argue that they are unable to attract sufficient deposits to fund growing demand for loans by rural businesses. Giving community banks greater access to GSE funds is one proposal. The FCS and the Federal Home Loan Bank system have both been raised as possible sources of funds for community banks (Duncan and others 1995; Barry and Ellinger 1997).

A related issue will be the possible development of rural secondary markets. Most observers agree that better secondary markets would be a welcome addition to the capital markets that serve agriculture and rural America. Farmer Mac is beginning to develop a stronger base of operation. Yet Farmer Mac currently can only securitize farm real estate and rural housing loans. A secondary market for rural and agricultural business loans would provide significant benefits, but there is no clear method for creating such a market. Many believe that a government sponsored enterprise (GSE) will likely be necessary for a secondary market for agricultural and rural business loans to succeed. But there is no agreement on which GSE is best suited to the task. Some observers support a "specialization" approach in which rural housing loans would be securitized by Fannie Mae and Freddie Mac, farm real estate loans would be securitized by Farmer Mac, and rural and agricultural business loans would be serviced by a newly created GSE (Vandell 1997). GSE status might be required to launch such a market, but in the long run the market might stand on its own.

Public policy steps that use GSEs to address agriculture's capital needs obviously cannot and should not be undertaken lightly, since they pose additional contingent liabilities for taxpayers. One of the major obstacles in considering an enlarged public role in agricultural and rural capital markets is that broad policy goals for federal programs in rural America are not well defined. Thus, there is no clear objective to weigh against the added liability posed by any new initiatives with a GSE. In every case where public policy has helped launch a GSE to fill a financial market gap, the new institution has been compatible with other broad public goals. Fannie Mae, for example, helped encourage home ownership, while Sallie Mae helped encourage more students to go to college. What is the public goal that any enlarged GSE presence in rural and agricultural capital markets will serve? There may well be one or more such goals, but they have not been clearly articulated by policymakers.

On the other hand, another factor to take into account in considering GSE involvement in agricultural capital markets is that the public already has a contingent liability in the form of the federal bank insurance fund. Community banks are an important source of credit for farm and rural businesses, and the public has stood behind the deposits that fund those loans. To some extent, therefore, implicit public support for GSEs is simply another form of contingent liability.

In short, both the ongoing contingent liability attached to the FCS and new questions about GSEs to serve rural borrowers will force policymakers to weigh public goals against public costs. When the FCS was formed in the early part of the twentieth century, the need for agricultural credit was plain, and the role of agriculture in public policy was fairly well-defined. Today, the credit needs of rural America are somewhat less apparent, rural goals are ill-defined, and federal expenditures are under great scrutiny. Thus, it seems likely that GSE involvement in rural America will expand slowly, if at all.

Assessing Relationship Risk

A recent trend suggests more agricultural production is occurring under various contractual relationships—for example, between input suppliers and producers or between producers and processors—and the winding down of government commodity programs may accelerate the trend. The trend is most notable in the rapid structural shift under way in the pork industry, which appears to be following the trail blazed by the broiler industry some 40 years ago. Similar structural shifts are emerging in the production of other farm commodities (Figure 3.3). In grain production, for example, some specialty grains are now grown under production contract to ensure the identity of the crop and that its unique qualities are maintained throughout the marketing channel. Thus, new marketing arrangements help ensure that farm products with unique, valuable attributes are recognized and rewarded in the marketplace.

Such contractual arrangements are also effective tools for spreading or sharing agriculture's traditional business risks, including weather-induced production problems or sudden shifts in volatile commodity prices. But new marketing arrangements may not eliminate these traditional risks. Instead, the industry's risks may only be transformed into a new kind of risk—relationship risk—arising from the chance that one or more parties to a business agreement will not perform as expected.

Relationship risk may be roughly proportional to the share of a firm's input or output that it plans to buy from or sell to another, and it may arise in various ways. For example, the party who has contracted with another for the production of a specialized product may be unable to finance the purchase of produc-

FIGURE 3.3 Production Under Contract or Vertical Integration,
1960 and 1993-1994

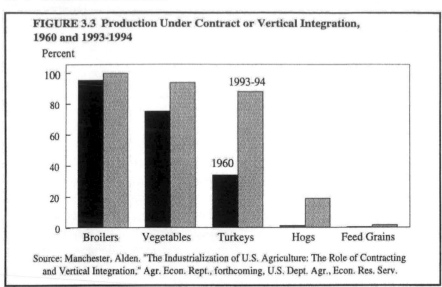

Source: Manchester, Alden. "The Industrialization of U.S. Agriculture: The Role of Contracting
and Vertical Integration," Agr. Econ. Rept., forthcoming, U.S. Dept. Agr., Econ. Res. Serv.

tion inputs as agreed in their contract. In that case, the producer may be forced
to seek financing elsewhere, perhaps at a high cost. Alternatively, the contractor
may be unable to purchase the completed product as agreed, and the producer's
only alternative could be the sale of the product at a discount in a residual market.
The disappointing revenue from the broken sales contract could limit the
producer's ability to repay a production loan to a third-party lender, who had
based the credit decision on the expected performance of the contract.

These simple examples highlight two underlying causes of relationship risk.
First, the parties to a production and marketing agreement may be unfamiliar
with the financial intermediation role they are assuming in their business agreement.
They may be accustomed to evaluating and hedging against traditional
production and price risks, but they may lack experience in evaluating their business
partner's financial wherewithal to perform as agreed. Second, the exchange
of financial information between the contracting parties may be incomplete, further
limiting one or both parties from fully assessing relationship risk.

A traditional approach to reducing relationship risk would involve a skilled
lender as the financial intermediary for a production or marketing agreement
between the two parties. A lender who financed one of the parties would likely
require complete financial disclosure from both, thus reducing in advance the
odds of nonperformance on the agreement. Alternatively, a skilled intermediary—perhaps a lender with appropriate industry experience—could be retained
to gather and review financial information from both parties. Thus, the intermediary
would broker the financial transaction, without assuming a financial stake
beyond a specified fee. The arrangement suggests the possibility of a lender
offering such risk assessment services to generate additional fee income. Taking

the approach a step further, the lender's fees might be calculated as a percentage of the income generated by the transaction for one or both contracting parties, giving the lender an equity-like financial stake in the venture. Thus, a regulatory concern could arise from the potential conflict of interest if the vendor of risk assessment services were also a lender to one of the contracting parties. In a similar approach that would resolve this concern, a consortium of lenders might staff a central office to offer comprehensive risk-assessment services to consortium members or others. This office would provide the necessary expertise to evaluate the financial risks of major contractors. Moreover, this information could be shared with regulators and thus promote a common understanding of risks in the sector.

Conclusion

The agricultural credit marketplace has always been dynamic, responding to cycles in the farm economy. Similarly, the regulations that set the rules of the game for agricultural lenders have undergone some changes in scope and degree over time. Today, the industry is becoming even more competitive, as new information technology renders obsolete old geographic and institutional barriers between lenders and as modest growth in credit demand is insufficient to accommodate a longer slate of ever stronger lenders.

The competitive marketplace is putting pressure on regulators to take a fresh look at lending regulations to ensure that regulatory goals are achieved with minimal regulatory burden. Costs and benefits of commercial bank regulations are being weighed anew, with the prospect that small community banks may well realize some easing of their regulatory burden. The entrance of large international banks will prompt a closer look at whether the financial playing field is level across countries. The Farm Credit Administration has recently completed a revision of rules governing the scope of FCS lending, yet the core issue of public liability for a single-sector GSE remains. Rising interest in enhancing rural capital markets, including the possible expansion of secondary markets, will focus still more attention on the role of GSEs in rural America, but a compelling case for an expanded role has not yet been made. Finally, all lenders find it difficult to assess the risk of lending to contract producers, and pooled efforts to gauge that risk may emerge in the future.

References

Ahrendsen, Bruce L., Alan D. Barkema, and Cole R. Gustafson. 1995. "Weighing regulatory costs in rural lending." *American Journal of Agricultural Economics*, 77 (3): 751-756.

Barry, Peter J., and Paul N. Ellinger. 1997. "Liquidity and Competition in Rural Credit Markets," chapter 3 in *Financing Rural America*, Federal Reserve Bank of Kansas City, pp. 47-78.

Berger, Allen N., Anthony Saunders, Joseph M. Scalise, and Gregory F. Udell. 1997. "The Effects of Bank Mergers and Acquisitions on Small Business Lending," mimeograph, Board of Governors of the Federal Reserve, May.

Duncan, Marvin, William Fischer, and Richard Taylor. 1995. "New Tools for Commercial Banks," *Journal of Agricultural Lending*, Spring, pp. 8-13.

Economic Research Service. 1997. *Credit in Rural America*, Agricultural Economic Report Number 749, U.S. Department of Agriculture, Washington, DC.

Farm Credit Administration. 1997. *Principal Differences Among Existing, Reproposed, and Final Capital Adequacy and Customer Eligibility Regulations.* McLean, Virginia, March.

Farm Credit Administration. 1997b. *Quarterly Report: Risk Analysis of Farm Credit System Operations, Quarter Ending March 31.* McLean, Virginia.

Federal Financial Institutions Examinations Council. 1992. *Study on regulatory burden.*

Gilbert, R. Alton. 1997. "Implications of Banking Consolidation for the Financing of Rural America," chapter 5 in *Financing Rural America*, Federal Reserve Bank of Kansas City, pp. 131-140.

Hoenig, Thomas M. 1996. "Rethinking Financial Regulation," *Economic Review*, Federal Reserve Bank of Kansas City, Second Quarter, pp. 5-13.

Keeton, William R. 1996. "Do Bank Mergers Reduce Lending to Businesses and Farmers?" *Economic Review*, Federal Reserve Bank of Kansas City, 81 (3): 63-75.

Spong, Kenneth. 1990. *Banking Regulation: Its Purposes, Implementation, and Effects.* Fourth Edition. Federal Reserve Bank of Kansas City.

Stam, Jerome M., Steven R. Koenig, and George B. Wallace. 1995. *Life Insurance Company Mortgage Lending to U.S. Agriculture.* Agricultural Economics Report No. 725, U.S. Dept. Agr., Econ. Res. Serv., Washington, DC.

Vandell, Kerry D. 1997. "Improving Secondary Markets in Rural America," *Financing Rural America*, Federal Reserve Bank of Kansas City, pp. 85-120.

Walraven, Nicholas. 1997. "Small Business Lending by Banks Involved in Mergers," mimeograph, Board of Governors of the Federal Reserve, March.

4

The Emerging Agricultural Lending System

Michael D. Boehlje

This chapter discusses the major changes occurring in the agricultural credit markets and in the systems/methods/procedures being used to provide credit to farm and agribusiness borrowers. The purposes of this discussion are to: (1) explain the dynamics of the markets and the changing role of various participants in those markets, and (2) provide an overview of the opportunities and challenges these changes provide to lenders in their planning and strategic positioning (the focus of the next chapter). I first discuss the new lending environment for agriculture and the new competitors in the market. Then I turn to the increased focus on the customer and market segmentation, the types of agricultural loans, and new delivery alternatives. Finally, I review other challenges, including sourcing funds, managing risks, reducing costs, pricing loans and developing alliances.

The New Lending Environment

Financing farm and agribusiness firms, particularly those in rural communities, will become an increasingly complex business venture for lenders. The financing needs and uses of funds by such firms are different today than in the past, and require new lending policies and procedures. And changes in firm size, structure, organization, and ways of doing business will challenge the traditional methods used to finance this sector.

Structural Changes

The agricultural and food production and distribution system will continue to undergo major structural changes during the next decade. The implications of these changes for financing needs and services are significant.

Changes in the market will encourage farm and agribusiness firms to seek alliances and partnerships at numerous levels. These partnerships may be among the input manufacturer, the producer, an integrator, and a distributor/dealer network with a strong presence in local markets. Alliances are likely to extend across the food chain, as in the hog industry where joint ventures among feed companies, genetic companies, integrators or producers, packing plants, and food retailers are in place or being discussed.

Increased integration and coordination of the food system has significant implications for lenders. First and foremost, the lender will need to have a more thorough understanding of a broader set of economic functions and activities to make solid credit judgments. Furthermore, an integrated food system will mean that individual firms may require financing for a broader set of activities, thus increasing the potential amount of funds needed as well as the types of financing terms and arrangements. If the coordination is accomplished through contracts or strategic alliances, the loan officer must not only evaluate the internal creditworthiness of the customer, but also the terms and conditions of the contract or alliance to make sure that the overall arrangement is financially sound.

For example, a contract grower may be financially sound, but the success of the venture may also be dependent upon the performance of a partner in the food chain such as a processor or retailer. This, the financial strength, the market position and the overall capacity of that partner to perform will be as important to the credit decision as the immediate customer.

Various coordination systems will likely change the asset structure or the cost and revenue characteristics of a firm, and thus require different criteria for credit assessment. For example, out-sourcing of supplies or contract production frequently reduces the current assets and liquidity of the firm since others own the inventory. Thus, typical measures of liquidity used in credit analysis—such as the current ratio or liquid asset margin—will need to be modified.

In contract production and strategic alliances, the terms of the contract and the sharing of risk between the various parties will be critical to the credit decision. In general, coordination and integration in the food system beget more interrelationships between firms and segments of the industry; sound credit decisions will require more knowledge not only about the firm, but about the related firms and the relationships between them.

Product Changes

Consumers have grown more demanding in the last decade. They expect quality control and products with specific characteristics on demand. High fixed costs at all stages of processing provide strong incentive to improve consistency of products flowing through the system. Plants and animals bred or engineered for specific end uses also will require production practices tuned to the complex and unique characteristics of the product.

Changes in the product characteristics in agriculture—from commodities to specific-attribute raw materials—and the pressures and incentives for processors to preserve product identity throughout the food chain have significant implications for farm and agribusiness lenders. Lenders as well as their customers must have a thorough understanding of the target or niche market for the product and the ability of the firm to exploit that niche (as well as exit from that niche market if it disappears). So marketing savvy, distribution skills and capacities, and manufacturing and logistics will all be important to the successful participant in markets characterized by increased diversity and product identity.

Some manufacturing and distribution processes associated with specialty food products may require very specialized equipment and facilities. The ability of the lender to analyze the risk of acquiring such collateral/security if the borrower defaults also will be important in lending decisions. If specific-attribute raw materials are needed for efficient processing or satisfying particular customer preferences, the assured supply of those raw materials will be critical to the overall financial performance of the processor.

Thus, the advantages of insourcing such supplies (vs. outsourcing with specific contractual and other arrangements) must be analyzed to project financial performance and assess the risk of loan default if adequate raw materials are not available. This is in contrast with the more traditional sourcing of raw materials from commodity markets, which rarely suffer from true shortages. Much of this identity preservation demanded by the marketplace will occur through strategic alliances rather than outright ownership, so the financial strength and performance of alliance partners will be important in assessing the creditworthiness of a particular firm within the processing chain.

Soft Assets/Hard Assets

Significant changes are occurring in the asset structure of farm and agribusiness firms, much like those changes occurring in the entire corporate sector. With the increased role that knowledge and information will play in the market position and financial success of such firms, large investments will be made in the soft assets of information, research and development, human resources and an organizational structure that is nimble and responsive to changing consumer demands and environmental conditions.

More hard assets—including machinery, equipment, and facilities—are likely to be obtained through leasing agreements, joint ventures, and strategic alliances where the firm acquires the services without the costs (or benefits) of ownership. These trends are evident in the pork sector in the form of contract production among producers, packers, and retailers; the increased use of leasing arrangements to acquire machinery and equipment in crop farming; and relationships between component suppliers and manufacturers/assemblers in the machinery and equipment business. In essence, the farm and agribusiness firm of the future will include more soft assets and fewer hard assets for two fundamental reasons: (1) the return on investment in soft assets will increase and be greater than returns on hard assets, and (2) the services of hard assets can be obtained in ways (i.e., leasing, contracting) other than ownership.

The implications of this trend to more soft assets are profound for agricultural lenders. First, the financial performance of agricultural firms will become increasingly dependent upon management and returns to management rather than ownership of assets and the capital earnings of these assets. Management will entail not only operations and marketing skills internal to the firm, but also successful negotiation of linkages with suppliers and distributors and having the proper external partners.

The classic approach to securitizing loans using hard assets will become increasingly difficult given the changing mix of assets within the firm. Lenders have yet to determine how to securitize loans with soft assets that are difficult to "attach or repossess" compared with hard assets. A further complication for the lending community of this relative shift to soft assets is that more of the cash flow and loan proceeds of the business will be needed to invest in and support those soft assets. Traditionally, money spent on human resources (research and development or training, education, and skill building) was considered an expense; in the firm of the future, it may need to be seen as an investment from the lender's perspective.

Boundaryless Firms

Firms traditionally included human, financial, and physical resources with clearly defined managerial control and delineation of financial performance/risk. This is no longer the standard in much of the industrial and service industries or in agribusiness.

It is not uncommon for many farms to lease equipment, but an increasing number of firms are contracting with others for such basic functions as chemical application, marketing, and agronomic decisions. Human resources (such as accounting and legal) are increasingly acquired from service industries, and on a consulting basis rather than through permanent employees. Increasingly, specialized input purchasing, marketing services, and strategic planning are being performed by "external" consultants.

Not only is the definition of a firm (in terms of the human, financial, and physical resources it owns) changing, but changes are occurring as well in the permanency of business focus and/or activity. Successful firms are becoming more dynamic and nimble — moving rapidly into and out of market niches; the business activities or the "products" of the firm change more frequently over time.

The boundaryless firm presents a unique challenge for lenders, particularly those that have worked primarily with traditional firms that have permanency/longevity, are led by career management with logical management succession plans, and include a substantial hard asset base. The boundaryless firm has few of these attributes; the "firm" and its activities (projects) may change drastically within two to three years. Key management expertise may be acquired externally through consultants or service companies. And many of the assets may be sourced through leasing, contracts, or other arrangements from other firms.

This type of firm is primarily a soft-asset, "deal making" entity, and has few of the traditional characteristics lenders value in a credit relationship. Yet the financial performance may actually be better, particularly when measured in such standard ways as return on investment or assets, than that of the traditionally defined firm because of increased nimbleness and liquidity. So the performance of a boundaryless firm may be enhanced at the expense of permanency and security, which are important from a credit perspective.

New Competitors

The role of various farm lenders has changed over the years (Table 4.1). The Farm Credit System (FCS) market share grew significantly during the 1970s and surpassed that of commercial banks, but the financial crisis of the late 1970s and 1980s impacted their portfolio more severely and commercial banks again became the dominant lender by the late 1980s. Commercial banks have histori-

TABLE 4.1 U.S. Farm Business Debt as of December 31, Selected Years, 1984-1996

Lender	1984	1988	1992	1994	1996
	Percent				
Farm Real Estate Debt					
Farm Credit System	43.7	36.6	33.7	31.7	31.5
Commercial Banks	9.0	18.5	24.9	27.1	28.6
Life Insurance Companies	11.1	11.6	11.6	11.6	11.6
Farm Service Agency[a]	8.9	11.5	8.5	7.0	5.7
CCC Storage Facility	0.6	0.1	0	0	0
Individuals & Others	26.6	21.7	21.3	22.6	22.6
Total ($ billion)	106,697	77,833	75,421	77,642	82,724
Non-Real Estate Debt					
Farm Credit System	20.8	14.2	16.3	16.2	18.5
Commercial Banks	43.2	45.9	51.7	53.1	51.6
Farm Service Agency[a]	15.8	20.9	11.2	8.7	6.5
Individuals & Others	20.2	19.0	20.8	22.0	23.4
Total ($ billion)	87,091	61,734	63,613	69,128	74,799

[a] Farmers Home Administration (FmHA) prior to October 1994.
Source: U.S. Department of Agriculture. *Agricultural Income and Finance: Situation and Outlook Report.* Economic Research Service, AIS-65, June 1997, Washington, DC.

cally focused on and dominated the non-real estate debt market, but much of the growth in recent times has been in real estate lending, where their market share is now almost equal to that of the FCS. Insurance companies' market share of real estate debt has declined from their dominant role to about 11 percent of the market today, while government lenders now account for a lower market share in both the real estate and non-real estate markets than in the past (Moss et al. 1997).

Noninstitutional lenders are important to the farm sector, accounting for 22 and 23 percent of the non-real estate and real estate markets, respectively, in 1996. This market share has remained remarkably stable over the last decade, but the types of lending, particularly in the non-real estate market, have changed. Captive finance companies (CFCs) have become much more prominent in recent times with loan volumes growing rapidly; it is estimated that manufacturers and dealers now have a 25-percent market share of the intermediate-term, non-real estate debt market for commercial farmers (Dodson 1997). And financial leasing arrangements are also growing rapidly; estimates are that one-fifth of commercial crop farms lease machinery and equipment.

Captive financing companies have been formed for many reasons, including expanding (or maintaining) market share in input markets, perceived conservative lending policies of traditional lenders, perceived profit opportunities on the part of input suppliers, and increased discipline in making credit decisions compared with traditional trade credit. Since the expansion of CFC programs in agricultural lending is driven by more than just excess capacity and competitive responses, such programs are not likely to disappear in the near or distant future.

This growth in CFC participation in agricultural lending has two major implications for the lending community. First, such firms will need to source funds. If those funds are sourced in the commercial paper market, through securitization in the secondary market or from the parent company through their own credit arrangements, such sourcing is expected to result in decreased lending activity for traditional or local lenders in the farm and agribusiness markets. If, in contrast, these captive finance companies source their funds through traditional credit lines, they may provide new business volume. Certainly, a significant opportunity for those institutions with access to national money markets would be to source funds for CFCs, which would in turn lend those funds to purchasers within the entire food distribution chain. This might not only include the financing of input supply firms that sell to farmers, but also floor planning and/or financing of merchant and open-account credit transactions between buyers and sellers within any part of the agricultural production and food distribution system. This strategy is being used by some banks and FCS lenders (for example, the merchant desk of Northwest Farm Credit Services) to compete with captive finance companies (see chapter 5).

A second major implication of the potential growth in CFC participation in the credit markets is increased competition with institutional lenders. The unique financing that captive finance companies emphasize—along with relatively efficient origination, servicing, and collection procedures—frequently enables them to provide credit services at an equal or lower cost and with more convenience than traditional lenders. Thus, such companies may be formidable competitors in the farm financial markets.

In addition to captive finance companies, new competitors in some markets include credit unions as well as leasing companies. Some specialized input supply companies such as Ag Services of America are providing financing as well. In the machinery, equipment and facility market in particular, leasing companies affiliated with manufacturers as well as independent companies such as Farm Credit Leasing and Telmark and others, are expanding volume at double digit rates. Although leasing historically had tax advantages for some producers under tax laws of the 1970's and early 1980's, many of those advantages have disappeared in recent times and the attractiveness of leasing is now more a function of down payment, interest rate and repayment terms. Given that the volume in the agricultural credit market has been growing at a modest rate of 1-3 percent per year in recent times, the double digit growth in lease financing along with the growth in volume of captive finance company financing has come at the expense of the traditional lenders. And the efficiency and customer attractiveness of these alternative sources of credit suggests that borrowers will increase their use of credit from these sources.

Customers and Market Segmentation

The 1970s was a period of rapid growth in loan demand in the agricultural credit market. Most lenders used passive marketing strategies, as suited the high demand growth: an "over the transom" marketing strategy—where lenders choose their customer base from those who walk through the doors—works well in times of excess demand.

But today's credit market, compared with the 1970s and early 1980s, is characterized by weaker demand, lower volume, and increased competition. In this environment, a much more aggressive marketing strategy is necessary to maintain market share and be a cost-effective supplier of agricultural credit.

Customer Needs and Relationships

Customer-focused agricultural lending requires an understanding of customers' wants and needs. Many lenders, when asked to describe their product or service from their customers' perspective, struggle mightily. They can describe the product from their viewpoint (an operating loan at 9-percent interest secured by inventories and machinery/equipment), but can't describe the features and benefits that the customer values, or why the customer does business with their institution rather than a competitor.

Understanding the customer, segmenting the market to better respond to their expectations, positioning the proper product to meet the customers' needs — in general, developing a marketing strategy focused on the unique characteristics and behaviors of individual customers—is critical to success in the increasingly competitive agricultural lending market.

Increased customer awareness requires that money be seen as a commodity. A loan product, in terms of its interest rate, repayment terms, and collateral requirements, is not all that different from a competitor's loan product. And furthermore, price is no longer a sustainable competitive advantage. Certainly, competitive pricing is essential to being a player in the market, but it is only "table-stakes"—it is not the basis to win and keep customers. If one lender lowers interest rates to attract or retain a customer, competitors will likely match that rate relatively quickly. This bidding war frequently ends with the lender that wins the business losing in terms of profitability because the loan is priced at an unacceptable margin.

Understanding why a customer does business with you beyond the desire for a competitively priced "commodity" loan product is critical to maintaining a customer base. Most borrowers indicate that the features and benefits of a loan product most valued are the convenience of obtaining that product, the speed of making a decision, the expertise and knowledge of the lender concerning broader issues of financial and business management, the information available to improve the borrower's decisionmaking, the spectrum of financial products and services available, and numerous other characteristics of the lending organization beyond the loan product. It is important to recognize that, particularly in service industries like agricultural lending, customers are buying experiences rather than products and services. Understanding the experiences a customer wants is critical to servicing the agricultural loan customer.

This focus on the customer is enhanced if the characteristics of the borrower-lender relationship are apparent to both parties. Figure 4.1 illustrates possible dimensions of a relationship; the dimensions identified suggest a continuum, for example, from a loan customer only to a customer who uses a broad spectrum of the institution's total financial products and services. The characteristics of a customer relationship that create value must be identified from both the lender's and the customer's viewpoint so as to best position financial products and services to fulfill customer expectations and generate profits for the lender.

Moss et al. (1997) have analyzed the impact of bank product and service offerings on competitiveness and the determinants of customer loyalty. Products and services that lenders judged to be important to borrowers were long-term service from the same loan officer; pre-set lines of credit; fixed-rate real estate loans without prepayment penalties; interest rate adjustments as a function of credit risk, loan size and term; and financial management and consulting services. As to customer loyalty; 50 percent of the survey respondents rated their farm real estate customers as highly loyal and 45 percent as moderately loyal; 76 percent

---

FIGURE 4.1 Dimensions of Borrower-Lender Relationships

Loan . . . . .Financial services
Personal . . . . .Business
Fickle . . . . .Loyal
Transitory . . . . .Lifetime
High touch . . . . .High tech
Core segment . . . . .Low priority segment
Antagonistic . . . . .Collaborative
Traditional . . . . .Industrialized
Advisor . . . . .Salesperson
Decisive . . . . .Needs hand-holding
Compatible objectives . . . . .Incompatible objectives
Equitable sharing of rewards . . . . .Inequitable sharing

---

of their non-real estate customers were rated highly loyal and 22 percent moderately loyal (Moss et al. 1997). The major factors impacting customer loyalty (as assessed by the banker respondents) in order of importance were relationships with loan officer, staff's knowledge of agriculture, stability of institution, stability of staff, rural inclination of bank, and levels of interest rates and services offered.

Market Segmentation

It is difficult for a lender to be "all things to all people." Attempting to provide a wide variety of products and services to any and all potential buyers, in essence, forces the lender to compete with all potential providers and negates any competitive advantage the lender may have.

Increasingly, agricultural lenders are recognizing the importance of developing a marketing strategy that recognizes differences in customers and targets specific customer segments that can be efficiently and profitably served. Developing a customer-driven marketing strategy requires answers to two critical questions:

(1) Who are our customers and what do they want? and
(2) Do I want them?

Answering the first question requires more than just a general awareness of how to anticipate and respond to customers' needs. It requires an assessment of the different kinds of customers that might be served—the market segments that might be pursued.

Different criteria might be used in market segmentation. Historically, agricultural lenders have segmented the market based on location (that customer is not within my trade territory and consequently will be difficult to serve); type of enterprise (we focus on hog producers rather than cattle feeders because we understand their business and they are the dominant type of producer in our market); size (our expertise and lending limit allow us to best serve customers with credit needs of $250,000 or less); or risk (we want only those loans that score five or better on our six-point credit scoring system).

Moss et al. (1997) use other criteria to identify three potential market segments: (1) large, commercial-scale farms, (2) small, part-time, or recreational farms, and (3) larger industrialized units involved in some degree of vertical coordination. Thus, new dimensions are becoming important in customer segmentation. Ways of doing business may become an important criteria for segmentation in the future. Independent producers have relatively traditional credit needs and can be evaluated with familiar credit and risk analysis tools and techniques. In contrast, integrated and industrialized producers may have not only different credit needs (for example, they may need funds to expand a distribution system or build a processing plant), but the traditional criteria for evaluating liquidity, solvency, and other credit risk may be impractical. Psychographic analysis, which evaluates the specific buying behavior of the customer and what drives his/her buying decisions (product attributes, quality, price, service, brand, information, convenience) is now being used by progressive agricultural lenders to better understand customer decisions and segment the agricultural loan market.

One approach to market segmentation attempts to describe four customer segments based on customer needs, collateral and risk, cost, and other characteristics (Table 4.2). These four segments are only illustrative (some would legitimately argue with the cost and other characteristics identified), and buying behavior or a psychographic analysis is an important missing component of the segmentation process. Still, to better understand and serve customers, it is important to recognize that different customers have different characteristics, need different products, and are attracted to a lending institution by different features and benefits. It will become increasingly important to target segments in a market that should be aggressively pursued in advertising and calling programs, vesus those who can be passively served.

A concrete answer to the second key marketing question—Do I want them (specific customers)?—requires definite answers to the following seven questions.

(1) Should this customer segment be a target? Given the market potential and my firm's ability to deliver products consistent with that market segment's wants (i.e., to sell the experiences they want to buy), should I focus marketing resources on this segment?

(2) How will I identify specific customers? To capture a particular segment of the market, you have to identify the specific customers and develop a unique strategy to close the sale for each of those customers.

(3) What specific products and services does this segment need? At this point, it is necessary to identify a product or service in very specific form, includ-

TABLE 4.2 Customer Segmentation

| | Loan Market Segment | | | |
	Traditional	Industrialized/ Integrated Grower	Small-Scale Producer/ Consumer	Investor
Credit Needs	Operating	Operating	Operating	Investment capital
	Modernization Incremental expansion	New venture expansion	Limited capital expenditure	
Size	$75-100,000	$250,000+	$25,000±	$250,000+
Market Potential	Declining volume and numbers	Growing volume and numbers	Growing volume and numbers	Growing volume and numbers
Risks	Traditional/ manageable credit risks	Market Technological Construction/ Startup Operating Regulatory Upgraded Managers	Traditional credit risks Nonfarm income?	Low repayment risk Other income
Collateral Vulnerabilities	Dated technology Environmental hazard	Environmental hazard Soft assets vs. hard assets	Small scale Dated technology Environmental hazard	Quality collateral
Cost (basis points)	200±	150±	250± (traditional delivery); 100+ (asset-based delivery)	100±

ing its features and benefits, rather than the generic form of operating loans, real estate loans, insurance services, etc.

(4) How will I deliver this product/service? Different products and services can be delivered with different distribution channels (by the loan officer, another bank employee, or an alliance partner on contract), and the distribution channel will have different costs and other characteristics.

(5) How will I price? Pricing of any financial product or service will depend on the cost of providing that service, competitor pricing, desired profit margins, and customer's price responsiveness or sensitivity.

(6) What credit standards or other qualifications will be used? Some customers can be evaluated with credit scoring models, and others might require more detailed analysis by a loan officer. Environmental audits might be necessary for some credits but not important for others. The potential volume of business that a particular customer can offer may not be adequate to provide investment counseling, trust management, or cash management services.

(7) What is the customer's lifetime value? The opportunity to sell core and related financial products and services to a customer segment over the lifetime of those customers may have a significant impact on the profit potential of a particular market segment.

Lifetime Customer Valuation

It is common to evaluate the earnings potential of a bond or a capital investment by assessing the flow of income that will occur during that investment's lifetime. Only recently have we come to recognize that we can and should view a customer and the income stream he/she will generate for the business over their lifetime in a similar fashion. The traditional approach to evaluating customer profitability has focused on the specific loan or other financial product or service transaction. But more recently, concepts of lifetime valuation have surfaced in the consumer goods market as well as the financial product and service market.

Assessing the lifetime value of a customer in the financial product/service sector requires two steps: (1) identifying the lifetime product chain, and (2) determining the lifetime revenue stream. In contrast to some of the more narrowly defined consumer goods industries such as electronics, apparel, and even automobiles, the financial products/service industries have unusually unique opportunities for selling a broad spectrum of products and services. This opportunity for cross-selling has been well understood in the financial community for years. The added feature of lifetime valuation is the explicit recognition of the changing mix of products and services needed by the customer over his/her lifetime, and the implications that larger lifetime revenue stream has for investments in customer solicitation and retention.

To illustrate the pragmatic application of lifetime valuation concepts, assume a financial institution can provide either directly or through some form of partnering or strategic alliance the financial products and services indicated in Figure 4.2. Given this list of potential products, the first step is to identify the set of products and services that specific customer segments might need during their lifetime—the lifetime product chain. An illustrative lifetime product chain is summarized in Figure 4.3 for a typical loan account customer. The second step is to combine this lifetime product chain information with cost and revenue estimates to generate a lifetime revenue stream. Table 4.3 illustrates a simple format that might be used for this estimation process. Note that revenues are gen-

---

FIGURE 4.2 Potential Financial Products/Services

---

1. Operating Loans
2. Capital Expenditure Loans
3. Deposit/Checking Services
4. Customized Credit Products
5. Joint/Consortium/Packaged Credits
6. Leases
7. Equity/Investment Banking Services
8. Property/Asset Management Services
9. Insurance/Risk Management Services
10. Cash Management Services
11. Retirement Investments

---

erated by the initial volume plus incremental growth in core products/service sales (in our case, the operating loan). But, over the customer's lifetime, revenues can be generated by the sale of related products and services, and the volume of these products may also grow over time.

With respect to cost, the initial upfront cost or investment is described as the courting or solicitation cost. In fact, these costs may be incurred for one or two years before core product sales and the resultant revenue. Once a core product is sold, closing costs are incurred and, in subsequent years, maintenance and updating costs will be required to maintain the relationship. In like fashion, the sale of related products will incur closing as well as maintenance and updating costs. The bottom line, as indicated in Table 4.3, is annual net revenue, which summed over time provides an estimate of the lifetime value of the customer.

FIGURE 4.3 Lifetime Product Chain

Life Stage	Product
Early Career	Deposit/Checking Services
	Operating Loan
	Capital Expenditure Loan
Mid Career	Cash Management
	Insurance
	Investments/CDs
Late Career/Retirement	Retirement Investments
	Property Management
	Trust Services

TABLE 4.3 Lifetime Revenue Stream

REVENUES	Time (years)						
	1	2	3	•	•	•	Total
Core Product/Service							
- Initial Volume							
- Incremental Growth							
Related Products							
- Initial Volume							
- Incremental Growth							
TOTAL							
COST	1	2	3	•	•	•	Total
Courting/Solicitation							
Core Product							
- Closing							
- Maintenance/Updating							
Related Products							
- Closing							
- Maintaining/Updating							
TOTAL							
NET REVENUE							

Estimating the lifetime value of a customer is not a trivial task—it requires cost and revenue information for the product and service line that might be provided by the financial institution, estimates of the probabilities of successfully selling the core product and related products, and the probability of retaining the customer core and related-product business. But the information on lifetime value is essential to understanding which customers have the highest potential payoff as the focal point of the lending institutions calling program, mail solicitation, and other market development activities.

Types of Agricultural Loans

Many agricultural loan officers have had painful experiences with the collateral-based lending of the 1970s and early 1980s. The focus of agricultural lending today is cash-flow lending. But do all farm and agribusiness loans need a cash flow? And in reality, is there a unique set of criteria that should be used for evaluating all agricultural loans? Answers to these questions are critical to cost-effective lending and strategic positioning in the rapidly changing and increasingly competitive agricultural credit markets.

Table 4.4 is one attempt to summarize the major types of loans made to producers. Analysis of customer segments and credit risks suggests that there are two categories of loans made to farmers (depending upon their size and type of operation): commercial loans and consumer loans. For moderate-size and larger full-time farmers, the primary source of income to repay operating or capital expenditure loans is the farming operation. In most important dimensions, such a loan is no different than that made to any commercial business venture and, thus, commercial lending practices adapted to the agricultural industry are appropriate for the evaluation of creditworthiness. For part-time farmers, where the primary source of income and repayment is off-farm employment, the principles and concepts of consumer lending, including detailed analysis of the level and stability of the off-farm job, are critical. In essence, consumer lending practices are more appropriate for this borrower than are commercial lending practices.

The type of loan also varies in the level of documentation required (Table 4.4). The first level defines the *signature* loan, where the financial strength of the borrower is so evident that further analysis (beyond a signature on the note) seems unnecessary. This type of loan is particularly appropriate for individuals with documented large net worths and/or annual incomes borrowing small sums of money, or those who, for example, may have sizeable certificates of deposit (CDs) or other investments in the financial institution and want to borrow only a modest sum.

The second level is the collateral or *asset-based* loan. Collateral-based lending has fallen in disfavor in agriculture in recent years, but under the right circum-

TABLE 4.4 Types of Agricultural Loans and Documentation Required

	Commercial	Consumer
Signature	Note	Note
Asset-Based	Note	Note
	Lien (priority)	Lien (priority)
	Financing Statement	Financing Statement
	Evidence of Marketable	Evidence of Marketable
	Collateral	Collateral
	(Liquid assets, insurance,	(Liquid assets, insurance,
	government program,	government program,
	contract or hedge, etc.)	contract or hedge, etc.)
Performance	Note	Note
	Balance Sheet	Balance Sheet
	Income Statement	Income Statement
	Cash Flow	Cash Flow
	Tax Return	Tax Return
	Lien and Financing Statement	Employment History
	Assignment of Equity	Off-farm Income
		Lien and Financing
		Statement
		Assignment of Equity

stances, asset-based lending may still be appropriate. The key determinants of whether such lending will result in low levels of risk are:

(1) Is the collateral relatively liquid and easily marketable?
(2) Does the lender have a first-security position in the collateral?
(3) Is the cash flow generated by the collateralized asset or enterprise relatively certain? and
(4) Does the borrower have a record of regular repayment?

If the answers to these four questions are yes, then an asset-based loan will typically be a very low-risk loan. Examples include operating loans for seed, fertilizer, and chemicals where the lender has a first position in the crop and the farmer hedges or forward-contract prices and purchases crop insurance. Another example would be a loan for feed or livestock where the lender has a first position and the feeder has protected his margin using futures or options markets. A key advantage of the asset-based loan is that much of the detailed financial analysis and documentation is not needed and the loan review process can be substantially shortened, increasing the cost effectiveness of the lending activity.

The third type of loan, a *performance* loan, requires full financial analysis of the efficiency, business performance, liquidity, solvency, and profitability of the business. This type of loan will require complete financial statements, current and historical income statements, and documented cash-flow analysis. The payoff to time committed in detailed analysis of this loan category can be substantial since the risk of nonrecovery on performance loans is typically much higher than that on signature or asset-based loans.

By dividing the loan portfolio into the three categories of signature, asset-based ,and performance loans, the lender can more effectively allocate time and expense in loan documentation and simultaneously reduce the risk of making inappropriate credit decisions.

Delivery Alternatives

Changes and innovations in the origination, delivery, and collection of agricultural credits may be necessary for lenders to remain competitive in the future. The cost of origination and servicing for agricultural loans using the traditional "bricks and mortar" strategy of commercial banks and Farm Credit System lending institutions is estimated at 175-225 basis points (Ellinger and Barry 1991). A significant portion of these costs (personnel, legal documentation, etc.) are invariant with loan size. Consequently, costs at these levels make it difficult to generate profits with smaller credits in the $30,000-$50,000 range.

A number of alternative origination and servicing alternatives are being tested in the agricultural credit market. One such option is the use of a portable credit card. Although the credit card arrangement may appear to be most appropriate for smaller volume purchases of supplies and parts, higher limits allow it to be used for major purchases of feed, fertilizer, chemicals, etc. Some input supply firms are using point of sale (POS) financing arrangements (e.g., Farmland Financial, Deere Credit Services, PHI Financial, etc.) for full-season financing of farmer purchases in contrast to traditional 30- or 60-day dealer credit. The efficiencies in origination costs subsidized by the product marketing activity, combined with specialized collection activity, result in POS delivery costs that are lower than those of traditional lenders by 50 percent or more (Henricksen and Boehlje 1995). This relative efficiency in credit extension, combined with the product marketing advantages of offering credit services, are major explanations for the substantial interest by input suppliers in expanding their role in the agricultural credit market.

The opportunity to combine credit card financing; a discount on product purchases (irrespective of outlet if they are purchased with the card); an accounting/recordkeeping service that summarizes by user-specified categories all expenses incurred using the card; and an annual summary useful for filing taxes and other government reports of all card transactions (as is done by Combined Rural Traders in New Zealand with their CRT card) may further expand the opportunities for POS delivery of credit.

Other innovations in credit delivery and collection include scorecarding and electronic telephone, Internet, or FAX origination and collection. Scorecarding, using limited information obtained from abbreviated loan application forms , has been regularly used by captive finance companies and is now being implemented by some traditional lenders for part of their portfolio. The typical score- card approach is to estimate the determinants of successful performance/repayment on agricultural loans from previous experience; and to use those determinants (in a mechanistic, simple-to-use scoring procedure) to triage the loan application into three categories: (1) accept, (2) reject, and (3) needs further analysis from a loan officer. This process is used to both improve the speed and lower the cost of credit decisions, and to direct loan officer time and energy to those credits that need more analysis.

Electronic delivery and collection includes loan-by-phone programs where credit customers dial an 800 number and answer a series of electronically delivered inquiries for information (maybe as few as 7-10 questions) with a preliminary decision in 15 minutes or less and all forms and legal documents delivered within 1-2 days by FAX or over-night mail. Internet and e-mail are also being used to communicate more rapidly with potential customers and to finalize loan agreements, as well as to communicate with customers in default to obtain commitments as to when payments will be made.

In summary, the role of computerized electronic credit evaluation and scoring techniques and targeted asset based lending under specified conditions are being used as ways of reducing delivery costs while controlling risks. Likewise, use of specialized collection services, including combining telephone contacts with trained collection field staff for direct contact are being implemented. The fundamental objective of these strategies is to provide cost-effective delivery of credit services while simultaneously controlling credit risks. Cost-effective, risk-controlled credit extension will be essential to maintain competitive interest rates for borrowers; increased competition in a slow growth market will dictate that lenders be more rate and service competitive than they may have been in the past.

Other Future Challenges

Given all the changes in the farm and agribusiness lending market, what are the most serious challenges agricultural lenders will face in the future?

Managing Risk

The financial stress of the agricultural sector during the 1980s increased lender's awareness of the importance of managing risk. Some lenders exited the market because agriculture was too risky an industry; others "repositioned" with a focus on certain kinds of agricultural loans. More emphasis has been placed on documenting and assessing risk-bearing ability and debt-servicing capacity of the customer. Also, risk assessment is now becoming a part of portfolio management and funds allocation decisions, as well as decisions concerning the credit-worthiness of individual customers.

Agricultural lenders have made significant changes in the last decade in their assessment and management of risk, and the dramatic decline in loan losses and even reversal of loss reserves is ample evidence of their success. The development and use of the recommendations of the Farm Financial Standards Council has improved financial documentation and enabled analysts to calculate profitability, efficiency, and repayment capacity measures, in addition to the traditional liquidity and solvency measures. The guidelines have also facilitated the development of databases and the availability of comparative data that can be used to assess the financial strengths and weaknesses of agricultural businesses. But new risks — relationship, regulatory and start-up risks, for example — are becoming more important. Agricultural lenders must broaden their perspective and expand their documentation to assess and manage these risks.

Reducing Costs

With the downsizing of agricultural portfolios during the late 1980s because of reduced credit demand and more conservative lending, cost control became a key element of profitable agricultural lending. For most lenders the reduced volume required paring back the staff and other resources allocated to the agricultural lending business. And some new competitors in the market — specifically captive finance companies and those who remained financially strong during the downsizing — have lower costs and price accordingly.

Cost control is essential for both competitive pricing and generating acceptable profit margins. Several information-gathering and credit delivery approaches are being used to reduce costs for these labor-intensive activities. As noted earlier the use of laptop computers to gather and analyze financial data while on the farm, loan-by-phone products, home pages on the Internet, and simplified application forms are examples of efforts to reduce the costs of collecting financial information. These user-friendly and convenient information collection procedures, combined with credit-scoring models, are a few of the efforts now used to reduce the costs of gathering information, analyzing that information, and delivering credit more cost effectively.

Sourcing Funds

For commercial banks in particular, the lower loan to deposit ratios of the past 10 years meant that obtaining adequate funds at reasonable cost to satisfy loan demand was not a serious problem. But loan to deposit ratios have been rising in recent times, and agricultural bankers are facing increasing competition from non-agricultural banks and other financial intermediaries in their traditional modest-cost deposit markets. No doubt most agricultural lenders will be able to obtain adequate funds to meet loan demand — the issue is will these funds have the cost and maturity characteristics that enable the lender to maturity-match loan demand and generate acceptable profit margins at acceptable risks.

The use of the secondary market for selling farm real estate loans has received only limited use by agricultural bankers. Legislation passed in 1996 changes Farmer Mac's statues and could lead to its increased use. The fact of the matter is that since Farmer Mac's inception, fund availability has not been a problem for many lenders. An increasing number of lenders may need to re-evaluate the Farmer Mac alternative in the months ahead. Private securitization, loan participations, and buying funds through jumbo CDs and other instruments may also be used to source funds, but must be managed judiciously.

Market Share and Pricing

Agricultural lending like any business is a margin business, and market share is essential to high performance. Increased competition has made it difficult for some agricultural lenders to maintain their volume of profitable loans. In essence, the lending capacity of traditional agricultural lenders did not shrink as rapidly as loan demand following the financial stress period of the 1980s, and in more recent times new lenders such as finance and leasing companies have entered the market. And it is not just a matter of market share — it is market share of the right kinds of customers that generate current profits and have the potential to general additional business and future profits.

Part of the challenge of maintaining market share is that of competitive pricing of loan products, but pricing must be adequate to generate a profit. The lower risk and most attractive customers are often the most rate sensitive. It used to be that lenders could price 150 basis points or more above their competitor without fear of losing most customers. In today's more competitive market, customers are well aware of a competitor's pricing policy and a lender probably has to be within 50 basis points of the competitor's rate. At the same, time, price is not a source of sustainable competitive advantage—if you lower your price, a competitor is likely to match it relatively quickly.

Pricing at a rate that adequately recognizes the risk of a loan is also critical. In too many cases higher risk loans are underpriced — the lender is not being adequately rewarded for the risk and potential loss he/she faces. Although the use of risk-adjusted rates has increased, there are still opportunities for improvement. A 1995 survey of agricultural banks in Illinois, Indiana, and Iowa (Moss, Barry, and Ellinger 1997) found that approximately 62 percent of commercial banks were using risk-adjusted interest rates for agricultural loans, with larger banks reporting greater use. That figure has increased from 47 percent found in a 1981 national survey conducted by Barry and Calvert (1981, 1983).

Developing Alliances

Agricultural lending in recent years has been primarily a singular and unique relationship between lender and borrower. Two or three decades ago joint ventures between lenders to serve particularly larger customers in the form of "correspondent banking" relationships were more common. With the growth in larger scale integrated and industrialized agricultural credits and the increasing demand for a broader set of financial products and services by agricultural customers including cash management, investment counseling, investment banking, etc., more joint ventures and alliances among lenders will be required. And increasingly the origination, servicing, funding and marketing activities in the agricultural loan market may be performed by different firms linked through alliances.

Furthermore, it is not uncommon with larger scale credits to use a lending consortium that includes a short-term lender to finance the facility construction with a take out agreement with the long-term lender; the same or a different short-term lender who provides inventory finance; an input supplier that guarantees part of the facility or operating loan or even provides some of the operating funds; and an investor who provides external equity capital. Picking the right partners in this circumstance, particularly when problems of debt servicing are encountered, is critical to success as an agricultural lender. And developing profitable alliances to provide related services such as asset management, insurance, investment counseling and estate planning will become increasingly important to being a high performance agricultural lender.

Conclusion

The purpose of this chapter has been to discuss some of the major changes that are occurring in the agricultural credit markets and in the methods and procedures that will be used to provide credit and financing in the future to farm and agribusinesses borrowers. These changes include the increased segmentation of the market based on demographics as well as psycographics; the changing competitiveness in the market, including increased competition among traditional lenders; and the entrance of aggressive new participants such as captive finance companies and leasing companies into the market.

Other changes and characteristics include more consortium lending and participation in joint or packaged financing rather than single source lending; increased financing of vertically coordinated food production distribution systems rather than single stages of production; increased performance based lending to soft asset based borrowers compared to collateral based lending to hard asset based companies; more sophisticated risk assessment and risk management strategies and requirements including explicit recognition of strategic as well as operational risk; the increased specialized nature of the asset base and the financial risk associated with financing those specialized assets; the increasing importance of strategic risk including the risk of changes in government policy, regulation and competitors response on customer's credit risk as well as the lender's operating performance and operating risk; increased emphasis on information as a resource and as a way to add value, manage risk, enhance profits, and capture market share; increased customer awareness of successful participants in the market, and the development of financial products and services responding to customer needs; a broader product/service mix demanded by customers including asset management, investment banking services, cash management services, etc.; increased consolidation of traditional lenders and more specialization in targeting financial products/services to selected customers rather than serving all potential borrowers; changing underwriting standards combined with computer based systems for credit-scoring and evaluation as well as real-time financial monitoring; development of multiple distribution/delivery systems tailored to unique borrower segments; increased separation of the origination, servicing and funding/financing of agricultural credits through securitization and secondary markets; the bundling of financial products and services with real products and services by various suppliers including captive finance companies, and the simultaneous unbundling of financial products and services by speciality companies such as leasing companies, financial consulting companies, etc.; and the changing nature of the human resources needed to service an increasingly complex, sophisticated agricultural credit market.

References

Barry, Peter J., and Jeffrey D. Calvert. 1983. "Loan Pricing and Profitability Analysis by Agricultural Banks," *Agricultural Finance Review*, Vol. 43, Ithaca, NY: Department of Agricultural Economics, Cornell Univ., pp. 21-29.

Barry, Peter J., and Jeffrey D. Calvert (1981). "Measuring Farm Lending by Commercial Banks in Illinois," 81-E-186. Indiana, IL: Department of Agricultural Economics, University of Illinois. August 1981.

Dodson, Charles. 1997. "Changing Agricultural Institutions and Markets: The Farm Credit Outlook." Presented at USDA's Agricultural Outlook Forum '97, Washington, DC.

Ellinger, Paul H., and Peter J. Barry. 1991. "Agricultural Credit Delivery Costs at Commercial Banks," *Agricultural Finance Review*, 51: 64-78.

Henricksen, Bill, and Michael Boehlje. 1995. "Cost Competitiveness and Profit Potential of Captive Finance Companies, *Agri-Finance*, June/July, pp. 50-53.

Moss, LeeAnn McEdwards, Peter J. Barry, and Paul N. Ellinger. 1997. "The Competitive Environment for Agricultural Bankers in the US." *Agribusiness* 13 (4): 431-444. John Wiley and Sons, Inc.

U.S. Department of Agriculture. 1997. *Agricultural Income and Finance: Situation and Outlook Report.* Economic Research Service, AIS-65, June, Washington, DC.

5

Emerging Strategies for Traditional Lenders

Allen M. Featherstone, Michael D. Boehlje,
and Joseph O. Arata

The face of agricultural lending is rapidly changing as the twenty-first century approaches. Traditional agricultural lenders need to develop strategies to deal with change or face the possibility of extinction. Some of the factors causing these changes include the internationalization of the U.S. economy, consolidation occurring in production agriculture, consolidation of the financial services industry, and the phasing out of farm program subsidies. Pressures from nontraditional lenders also continue to cause traditional lenders to change.

This chapter will discuss emerging strategies of traditional agricultural lenders. The major emphasis will be on factors needed to develop a strategic emphasis for traditional agricultural lenders. The current lending environment will first be examined as reflected by statements from individual lenders regarding their mission. Then, generic strategies for developing a strategic direction will be discussed in the context of agricultural lending. Also, the process of developing a strategic plan will be discussed. Next, the process of strategy formation and the concept of thinking strategically will be examined. The chapter will conclude with a discussion of specific strategies that traditional agricultural lenders might adopt.

The Current Industry

A significant part of the current agricultural lending industry can be characterized by relationship lending. Relationship lending implies a higher level of service than the alternative type of lending which is often called transaction lending. Transaction lending is based strictly on the principle of many transactions at the lowest possible cost.

The current U.S. agricultural lending industry is made up of many players. These include banks, the Farm Credit System, life insurance companies, the Farm Service Agency, farm suppliers, individuals, and most recently, international banks. To some degree, relationships are a focus of each of the major players.

The Farm Credit System is organized as a cooperative where the board of directors consists primarily of farmer borrowers who are often relationship-minded.

It is our vision to have an aggressive strategy with a dedication to sell our products and market our organization as the preferred agricultural lender in Northeast Kansas. All planning and relationships will be approached as an aggressive, caring journey. Conscious attention will be given to maintaining a sound, profitable, professionally-run business while taking sufficient risk to succeed.

—Farm Credit Services of Northeast Kansas, *1994 Annual Report*

Banks offer an alternative source of agricultural credit. The banking industry serving agriculture ranges from very small community banks to very large multiregional banks. While the corporate mission of these institutions will vary from bank to bank, service and cost are important components in the industry.

Norwest Financial distinguishes itself from its competitors in other important ways. Everyone in a Norwest Financial store sells products and serves customers. Everyone sets goals that are monitored daily, weekly, monthly and yearly. Compensation is based on reaching these goals. All credit decisions are made by people at the store level. Every customer is a sales opportunity. Expenses and staffing are lean. There are only five levels of management between the CEO and the customer.

— Norwest Corporation, *1996 Annual Report*

Agricultural lending has shifted recently due to the influx of what are often referred to as captive finance companies. These companies finance the sale of their product. The companies often raise capital using the same markets that they use to finance their manufacturing operations. Relationships and flexibility are also key elements in the lending process.

Case Credit has a legacy of customer service. Our customer relationships span generations. We're proud of that. Our experience, together with constant customer contact, is our strength. The knowledge of our customers' business enables us to create and structure financing and leasing programs that are tailored to their unique needs.

— Case Credit Web Page 6/16/97

Multinational banks are another competitor in the U.S. agricultural lending market. Two major players are Rabobank and Credit Agricole. However, the general focus of lending practices of these international players does not appear to deviate substantially from traditional U.S. lenders.

> Throughout our history, Rabobank has always been what is now called a 'relationship bank'. And for almost a century, our striving has been to provide products and services relevant to our clients.
>
> — The Rabobank Group, *Annual Report 1995*

Although it appears that numerous entities provide agricultural credit to a broad spectrum of customers, other entities are primed to offer credit under profitable arrangements to large producers. For example, investment bankers potentially could serve the needs of larger agricultural producers at a competitive cost.

While the focus of the agricultural lending market appears to be relationships and service features (while controlling cost), cost could become increasingly important. With continued advances made in the financial services industry, the future viability of current lenders may ebb and flow. Traditional lenders will need to adapt strategies that will allow them to continue to function, merge with other institutions, or cease business operations. Several generic strategies that may guide future business decisions are available to banks and other traditional lenders.

Generic Strategies

Numerous strategies exist to position a financial institution for the future, each of which prescribes an objective. Mintzberg (1991b) identifies several generic business strategies. Those relevant to the financial services industry include price differentiation, image differentiation, support differentiation, innovation differentiation, or a combination.

The price differentiation strategy involves providing a service at the lowest total cost. Price differentiation requires an aggressive pursuit of activities that will reduce cost, maintain the required margin, and allow the firm to charge the lowest interest rate for a loan. In the financial services industry, it may involve standardization such that only certain types of products are offered. Customers that are marginal credit risks or that desire nontraditional terms or collateral backing are not typically served. Price differentiation may require strict credit scoring models where a decision can be made mechanistically. Positioning as a cost leader does not allow one to totally ignore quality of service; however, the driving force in the management of the firm is cost control. While a credit scoring system may reduce costs, it may also allow better service—for example, reduced turnaround time to loan approval.

The image differentiation strategy involves creating differences where differences may not exist. This could involve cosmetic differences used to create an image. In the financial services industry, this strategy could be manifested in a "hometown" bank philosophy, or "farmer-owned" lending institutions. An underlying argument is that positive externalities may accrue to the local economy. Hometown banks add to a rural community's infrastructure by providing local jobs, adding to the tax base, and supporting local athletic and civic events. In addition to providing general and specific banking services, rural bankers are often community leaders who provide financial advice and expertise to elected officials and various charitable groups. Aggressive well-managed banks can be catalysts for regional entrepreneurs and community development.

Support differentiation, without any effect on the product itself, allows for differentiation based upon related services. The financial services industry may offer related services such as appraisal services, insurance services, or deposit services. Phrases like a "full service" bank indicate support differentiation, with services provided beyond the immediate product.

Innovation differentiation is labeled by Mintzberg (1991b) as design differentiation. Differentiation is based on the offering of an innovative product that other firms do not offer or harnessing new types of technologies to a current product. In the financial services industry, this has manifested itself in the early adopters of automated teller machines (ATMs), Internet banking, and the development of equity financing arrangements. Innovation differentiation is not keeping up with competitors, but being two steps ahead of them.

Most lenders will use a combination strategy, adopting some form of the multiple strategies noted above. Lenders may also choose alternate strategies based upon product line.

Developing a Competitive Advantage

In order to choose an appropriate strategy upon which a competitive advantage can be built, it is important to examine the competitive environment. The competitive environment consists not only of knowledge of others but of one's own business. To this end, it is critical that a financial institution have a good understanding of the profit associated with each of its product lines. Gilbert and Strebel (1991) identify the three steps to building that competitive advantage: industry definition, identification of possible competitive moves, and selecting among generic strategies.

Industry Identification

Gilbert and Strebel further argue that the definition of an industry occurs by identifying the boundaries of the industry, the other players in the industry, and by defining the rules of the game. Identifying the boundaries of the industry involves examining the entire chain of activities. In a lending situation, this involves examining not only the loan types and the related services but also the mechanism for obtaining funds. A financial institution may not have a competitive advantage in assessing credit risk, but it may have the ability to obtain funds at a lower cost or an ability to better manage and match assets and liabilities.

Examining the value chain in this manner may allow a financial institution to compete at one level by using a competitive advantage at a previous level. Captive finance companies essentially have extended the value chain by not only providing the product they sell—for example, machinery, feed, seed, and fertilizer—but by also providing the financing often needed to purchase the product.

Learning the rules of the game first involves the determination of what the customer expects with respect to quality of product and price. Charging a high price for what the customer perceives as a low-quality product will eventually result in business extinction. In addition to determining what the market dictates with regard to price and value, it is also imperative to determine the regulatory rules of the game. For example, in expanding up or down the value chain, are there government regulations that differ from those currently required, or do requirements differ from competitor to competitor?

Finally, all existing players should be identified and analyzed. Players involve both competitors and those other participants that perform vital activities in the industry. For example, several smaller machinery companies do not provide financing with their product. But these companies could be considered players in the lending industry in that strategic alliances could be built with these companies to offer financing arrangements. Competitors involve not only those currently in the industry but also those that could enter the industry.

Competitor Analysis

A competitor analysis seeks to ascertain how competitors may react to certain situations. This is not unlike the analysis of tendencies that occur in athletics. For example, in football on third down and three, what are the tendencies or chances of a run? In a declining interest rate market, what are the tendencies of a competing financial institution? The ability to predict the reaction of competitors under various situations can help determine the appropriateness of certain strategies. Porter (1980) suggests a number of factors to consider when analyzing a competitor; these include analyzing goals, attitude toward risk, market position, organizational structure, controls and incentives, leadership, composition of the board, accounting system, and affiliation with parent companies.

The first step of a competitor analysis involves analyzing the goals of each competitor, which indicate long-run objectives. Is the objective to maintain market share, a stable profit, or a combination? In addition to long-run goals, many firms have short-term goals. How are the long-run objectives traded off with short-term objectives? Answers to these questions can lead to an understanding of a competitors's retaliatory moves.

The ability of a competitor to deal with risk is also an important component to consider. Among lenders to agriculture, the ability to withstand risk varies dramatically from player to player. Some firms are industry lenders but with limited geographic diversification; others are national in scope. Some have agricultural lending as the primary focus, others as a minor focus. All of these characteristics suggest differences in risk absorption capacity and responses across institutions.

In addition to business goals, does the competitor have beliefs or values, economic or noneconomic, that affect firm decisionmaking? These beliefs or values are sometimes referred to as superordinate goals. They are guiding concepts or management's notions of future direction, and are often unwritten. For example, does the firm want to be considered a technology leader, a market leader, or an industry statesman? Does the firm have a tradition or recognizable patterns of reaction?

The organizational structure is important in analyzing a competitor. Among agricultural lenders, the forms of organization include closely held private corporations, publicly traded corporations, subsidiaries of parent corporations, farm-owned cooperatives with implied agency status, and government entities. The ability of these organizations to be innovative and to react in a timely manner varies from firm to firm. In addition, these organizations face differing regulatory environments; some are highly regulated while others face very little regulation.

Control and incentive systems are important in analyzing competitors. The incentives for a government agency will differ greatly from a publicly traded corporation. Accounting for differences in control and incentive systems may dictate different reactions to different competitors pursuing the same strategy. In addition, the background and experience of the individuals leading the competitor are important factors to consider. For example, the reaction of an individual who underwent the farm crisis of the 1980s may differ from the reaction of an individual who did not. In addition, the length of time in the industry causes different lenders to have differing institutional memories.

The composition of the board is also important to consider when analyzing a competitor. Is the board controlled primarily by insiders or is there outside representation? What occupational background does the outside representation have? Does the board have intimate knowledge of the industry at hand? For example, if a product line is a secondary or tertiary line, what type of expertise in that line of business is represented on the board? Is there general agreement among the board and the top management regarding the future direction of the firm? These issues can provide an indication of the ability and desire to weather adversity.

Another factor in assessing a competitor is the accounting system used by the firm. For example, when the Farm Credit System (FCS) used average costing of funds, it gave them a distinct competitive advantage during periods of increasing interest rates and put them at a distinct disadvantage during periods of declining rates. Other issues such as the allocation of cost among different product lines could affect a competitors' performance.

A number of agricultural lenders are subsidiaries of a parent company. This leads to issues in addition to those above. A subsidiary may be constrained by the parent company, depending upon the strategic importance placed on the subsidiary. Issues such as why the subsidiary was formed and whether it is a profit center or a loss leader are key in determining the likely response to competitive pressures.

A final step is trying to ascertain what the competitor believes about itself. A competitor could have unrealistic opinions about their place as a market leader or low-cost producer. Whether or not these opinions are true, the competitor will respond based upon their perceptions. It may also be useful to ascertain the perceptions a competitor has about your firm. This information may be useful in determining how strategic moves will be interpreted by a competitor.

Competitor Intelligence System

How does a firm find information to perform a competitor analysis? Information sources are both public and private. Public sources include annual reports, filings to government agencies (10-Ks), speeches by management, newspaper articles, business press, trade magazines, the Internet, and industry studies. Much of this information is available electronically which lowers the search cost dramatically. Private sources of information include consultants, trade associations, security analysts, former employees, professional meetings, and users of the product.

Competitive Moves

Competitive advantage is built around being able to provide the product that the consumer desires at the lowest cost. Gilbert and Strebel (1991) argue that the business systems and an industry's stage of development affect the competitive moves available to a firm. Maintaining a competitive advantage involves delivering higher quality at the same cost, or the same quality at a lower cost. In any event, the business system must provide superior performance at one point of the process. For example, a financial firm with superior asset/liability management may be able to offer agricultural loans at a lower cost than other lenders, even though it has no competitive advantage in making agricultural loans. The superior performance does not need to occur at the end of the value chain.

Identifying Industry Development

Before a business is able to build a strategic focus, it is essential to properly identify an industry's stage of development. Porter (1980) identifies three distinct industry environments: (1) emerging industries, (2) mature industries, and (3) declining industries. Just as inaccurately assessing the financial situation of a potential borrower causes duress to a financial institution, so does inaccurately assessing the characteristics of the industry. Correctly assessing the business environment is crucial to making strategic management decisions.

Emerging Industries

Porter (1980) argues that an emerging industry can either be newly formed or newly re-formed. Re-formed industries can be caused by changes in technology, emergence of new customer needs, or shifts in relative cost relationships. The major characteristic of an emerging industry are that the rules of the industry are undefined.

Porter argues that an emerging industry is also characterized by the lack of rules for competition. Much of this arises due to the uncertain nature regarding product specifications, service requirements, and product technology. Are there characteristics of an emerging industry that could be relevant to traditional lenders?

The production agriculture sector is currently facing more risk with the passage of the Federal Agricultural Improvement and Reform Act of 1996 (FAIR) (P.L. 104-127). The rules of the game have changed and farmers will need to manage this risk more effectively themselves or to purchase products that allow them to shift that risk to others. The risk balancing hypothesis of Gabriel and Baker (1980) is widely understood in agricultural finance circles. The basic tenet of this hypothesis is that farmers have some level of risk that they can and are willing to bear, and that financial and business responses occur to alleviate or mitigate risk. In response to the risk that has been shifted to the sector, financial or business responses such as increased use of insurance, diversification, contract production, and forward contracting are expected. Given the lack of a well-developed equity market, which to some degree limits financial responses, an emerging industry that could be of interest to traditional lenders is a risk management industry. This industry would allow farmers to transfer risk to others. However, the effect on traditional lenders and the degree of participation is uncertain. How farmers deal with this risk and the products that they will demand is uncertain. Will farmers adjust by using less leverage in relation to equity, which could shrink the market for farm debt; will farmers demand a product that will allow them to access equity markets; or will all of the adjustment take place within the operations of the business?

Recent legislative and regulatory changes in the banking industry may lead to the characteristics of an emerging industry, but it is important to note the background of these changes. Historically, three laws have significantly restricted the evolution of the financial services industry. The Pepper-McFadden Act of 1927 (P.L. 69-639) placed geographic restrictions on the ability of commercial banks to operate in more than one state. The Glass-Steagall Acts of 1932 and 1933 separated the commercial banking and the securities industries.

The Pepper-McFadden Act permitted national banks to establish full-service branches within the cities in which they were located if state banks had similar authority. Following the passage of the Pepper-McFadden Act, states began to increase restrictions on branch banking activities. The effect of the Pepper-McFadden Act was to prohibit interstate branching by national banks and state-chartered Federal Reserve member banks.

The Glass-Steagall Act of 1932 (P.L. 72-44) temporarily liberalized the nature of collateral against rediscountable paper and permitted the temporary use of obligations of the United States as collateral for Federal Reserve notes. But more importantly, the Act required separation of investment activities and commercial banking. The Banking Act of 1933 (P.L. 73-66), also known as the Glass-Steagall Act, was the first of the major banking laws enacted by the Roosevelt administration. It contained a number of major provisions, including: (1) the redefinition and continuation of restrictions on branch banking; (2) the requirement that member banks sell their security affiliates within one year, and the exclusion of securities firms from engaging in the banking business in the future; and (3) the prohibition of banks that generate numerous asset-based securities from underwriting such activities.

Banks, especially money center banks, challenged the latter prohibition, but the investment industry strongly supported the prohibition. In 1987, the Federal Reserve interpreted a section of Glass-Steagall to allow underwriting activities if they were conducted through a securities subsidiary "not principally engaged" in underwriting. This was interpreted to mean deriving less than 5 percent of revenues therefrom. A few banks were granted the opportunity to underwrite commercial paper, municipal reserve bonds, and asset-backed securities. Other provisions of the Glass-Steagall Act historically have also been opposed by the commercial banking industry.

Commercial bank consolidation has been ongoing for decades, but the pace of mergers has accelerated in the 1990s, and recent legislative changes imply that amalgamation likely will increase (LaDue and Duncan 1996). The Riegle-Neal Interstate and Branching Efficiency Act of 1994 (P.L. 103-328) established more uniform interstate branching across the nation. It allowed multibank holding companies (MBHCs) to acquire legally separate bank affiliates in any state beginning in 1995, and to convert those affiliates to interstate branches starting June 1, 1997. The post-June 1, 1997 interstate mergers must be between adequately capitalized and managed banks, subject to concentration limits, state laws, and Community Reinvestment Act (CRA) of 1977 (P.L. 95-128) evaluation by the Federal Reserve. The CRA was passed to further the congressional intent that banks meet the credit needs of their local communities (a type of affirmative action program for neighborhoods or communities) and to encourage investment in the immediate localities served by depository institutions.

States were given three years (1994-97) to opt out of interstate branching, and only two chose to do so. All states maintained the authority to control the extent of branching within the state. While the Glass-Steagall Act remains in force, regulatory changes now allow national banks and bank holding companies to own securities firms as subsidiaries, provided that the subsidiary obtains no more than 25 percent of its revenue from underwriting and dealing in securities. Securities firms are not yet allowed to buy banks.

To place recent developments in perspective, interstate banking in fact was fairly common by 1995, with MBHCs owning legally separate banking affiliates in various states. The Riegle-Neal Act's clause that became effective in 1995 extended this by allowing holding companies in any state to acquire bank affiliates in any other state. Previously, some states did not allow their banks to be acquired, or insisted that the acquiring company come from specified states. The Riegle-Neal Act's clause that became effective in 1997 legalized widespread interstate branching by allowing MBHCs to merge their affiliated banks as of June 1, 1997, effectively converting bank affiliates to branches. If an MBHC so desired, it could reorganize itself to own a single bank with branches in many states. In short, states have been given some latitude to stop, or at least slow, the movement of their banking structure toward national integration by imposing market-share limits and providing temporary protection to new banks.

There has been considerable movement toward removing geographic and other restrictions on commercial banking. In addition to the geographic deregulation, regulators have nibbled away at the boundaries for insurance and securities activities by using their authority to permit bank subsidiaries to perform activities closely related to banking. The result is more freely operating commercial banking markets. Large global money lenders now tend to assess how their customers score on their credit applications rather than where their customers live. Credit terms and interest charges now more likely will be based on a standardized analysis. Money can and will be transferred without handshakes or face-to-face conversations. Debt and equity instruments can both be used by a single entity to provide full-service financing rather than relying solely on a total debt or total equity instrument. Thus, the desire and the ability to provide new products are converging.

Mature Industries

As an emerging industry moves from a fast growth industry to more modest growth, the management strategies that a traditional lender will want to focus on will change. As the industry matures, Porter (1980) argues, it will be characterized by more competition for market share. Firms begin to face repeat buyers that have substantial experience with the product, competition places more emphasis on service, the addition of capacity becomes a more critical decision, new products or advances are more difficult to come by, international competition increases, and the profitability of the industry declines.

To a degree, many traditional agricultural lending functions are occurring in a mature market. There is stiff competition among lenders; the speed of making loans is crucial, international and nontraditional lenders abound, too many dollars are chasing too few loans, and profit margins are thinner. These characteristics of the current agricultural lending market may lead traditional lenders to exhibit behaviors that could jeopardize the competitive long-term strength of the firm.

Porter suggests several pitfalls that managers in a mature industry should avoid. These include unrealistic evaluation of their competitive position, being caught in the middle, having excess cash, giving up market share too easily in favor of shortrun profit, irrational reaction to price competition, overemphasis on "new" products at the expense of aggressively selling existing products, clinging to notions of higher product quality versus aggressively meeting the competition, and holding on to excess capacity. Many of these behaviors can be observed in the agricultural credit markets.

Declining Industries

As an industry begins to decline, there is a shift in the pressures that management faces. A declining industry is characterized by an absolute decline in sales over a period of time. Porter (1980) suggests that the process by which demand declines has important consequences for competitive behavior during the decline phase. He argues that uncertainty with regard to demand, the rate of decline, the structure of remaining demand pockets, and the cause of decline are important when considering the competitive strategy in a declining market. Firms that fail to recognize that demand is declining can cause significant disruptions in the normal process of decline; in an effort to reposition themselves for revitalization or to utilize excess capacity, they may cause price warfare. As all firms generally agree that an industry is in decline, exit strategies begin to be formulated that lead to a more orderly transition.

One characteristic of a declining industry that is particularly noteworthy in agricultural lending is a decline in the size of the customer group. Production agriculture continues to consolidate into an increasing number of smaller consumer-like credits, but a shrinking number of larger commercial/business-like credits. Thus, as competitors divide a smaller customer base, competition becomes more intense.

Identifying Strategic Groups

Lenders can be positioned into groups based upon the moves that they make at a given time. Groups may help in the examination of the life cycle of an industry. The interaction of the market and moves made by individual lenders can affect this life cycle. For example, a lender that is first to develop an equity product would have the advantage of defining the parameters of that product. This might be a pre-emptive strategy—a similar product of equal or better quality would be more difficult to develop. Identifying these groups also allows one to understand the competitive position of different firms.

Example—Feeder Cattle Financing

In developing an industry analysis and strategic grouping, it is important to define the product line to be examined. An individual lender could focus on specific elements of the agricultural production system in a geographic area; for example, a lender could focus on cow-calf producers and the financing of calves and/or feeders through the cattle feeding process in western Kansas. Farm operators who do not finish their own cattle generally do not benefit fully from improved genetics and the biological production process in which they have invested. Owning cattle through the feedlot (retained ownership) as opposed to selling them as calves or feeders requires additional capital and a longer ownership.

The competition for financing retained ownership will come from a number of industry participants. The FCS holds about 9 percent of agricultural non-real estate loans. The Farm Service Agency (FSA) has approximately 3 percent of non-real estate loans; it provides mostly guaranteed loan assistance for beginning farmers and ranchers. A comparative advantage a lender might consider is a specific loan program designed for FSA customers who otherwise are unable to obtain financing (to retain ownership) from commercial credit sources. Other sources of competition for financing retained ownership include individuals, life insurance companies, broker dealers, feed companies, the Small Business Administration, and foreign banks. These organizations have access to lower priced capital, but may not be able to compete as effectively in terms of customer service.

Figure 5.1 provides a picture of the current industry structure in terms of strategic groups. The horizontal axis represents a qualitative estimate of the service provided by each lender on a scale of 1 to 10. The vertical axis represents the price (interest rate) for a feeder cattle loan subtracted from the prime interest rate as of the summer 1997. The firms in the oval are considered traditional sources of credit. Financing, however, is available at much more favorable rates from nontraditional lenders, but with considerably reduced service. Forming an alliance between a traditional lender and broker-dealer could possibly provide lower funding costs and increased resources for a funds-constrained lender. However, funding costs for the traditional lender may not necessarily be less than those experienced by using a broker-dealer. A survey conducted in fall 1995 by Cole, Featherstone, and Albright (1996) found an average spread of 4.1 percent between the cost of funds and the interest charged on an agricultural loan. Eighty-five percent of Kansas banks had a spread between 4 and 5 percent. Thus, many Kansas banks have been able to obtain funds at rates lower than the

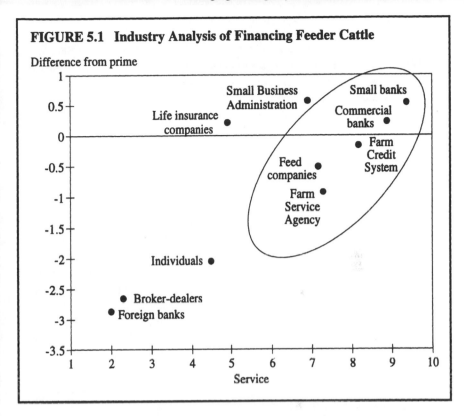

FIGURE 5.1 Industry Analysis of Financing Feeder Cattle

broker-dealer. However, the cost of funds for rural banks is rising as they must compete with higher rate savings instruments available from the national and international markets. Borrowers that do not want service could begin to bypass traditional lenders and take advantage of less expensive credit elsewhere.

Those lenders who emphasize service could develop an image differentiation from other lenders by providing customers with information about the risk and profit potential of finishing a specific year's calves or feeders. Information concerning the cattle situation and outlook from university/extension and/or private sources could also be provided.

The most effective comparative advantage a lender could have is to be able to provide funds to cattle producers at lower costs than the competition with slightly better service. Although there is currently no well-developed secondary market for cattle loans, there always exists the potential to combine a group of loans and offer them for sale to a large broker-dealer if available funds are constrained. Opposition to the sale of short-term cattle loans by lenders could be quickly overcome if part of the lower cost were passed along to them.

Strategy Formation

After examining the competitive landscape, the next step is to formulate strategy. Mintzberg (1991a) argues that development of strategy is more of a crafting process than a planning process. He suggests that strategies are both plans for the future and patterns from the past. They need not be deliberate—they can also more or less emerge, develop in all kinds of strange ways, and often happen in brief quantum leaps.

It is important to note that strategy involves both patterns from the past and plans for the future. Often strategy is thought of as only a forward-looking process. However, consistent patterns of past behavior often are used to formulate the strategic direction.

Strategy often is not formed in a deliberate process. Many times strategy develops as a reaction to certain events. A loan officer is trying to increase market share. The loan officer realizes that often loans are not being booked because of inflexible terms. The loan officer uses this information to develop simple spreadsheets and flexible payment plans that give the same return to the institution but are more accommodating to the customer. The lender eventually formalizes these ideas into a new line of products. In this case, the products were not conceived by a deliberate process—they were developed in response to outside pressures. In contrast, a credit scoring model could have been adopted by a deliberate formal process to reduce response time and lower cost.

Many times, strategy can come from the bottom up. Mintzberg suggests that effective strategies develop in many ways. Grassroots personal experimentation causes strategies to develop from the bottom up—individual ingenuity provides the engine for development. The setting of broad guidelines by management is more a top-down approach; however, the details are left to the organization. Finally, management may dictate a process for an organization to walk through in order to develop strategy. Each of these methods can develop effective strategies or a comparative advantage.

Implementing Strategy

Many strategies fail because of execution. Peters (1991) argues that execution is the key ingredient in determining whether a comparative advantage can be developed and maintained. The execution of a strategic plan must be carried out by competent employees, who are characterized by three traits: the desire for total customer satisfaction, the desire to continuously innovate, and the ability to treat others with respect. Each of these traits will lead to the development of a long-run competitive advantage.

The innovative lender is focused on the customer, actively seeks new ways of doing business, and expects that all employees will contribute to the success of the firm. Peters argues that market leaders have a passion for their product and are willing to take risks. He quotes George Gilder from *Wealth and Poverty*:

Economists who attempt to banish chance through the methods of rational management also banish the only source of human triumph. The inventor who never acts until statistics affirm his choice, the businessman who waits until the market is proven — all are doomed to mediocrity by their trust in a spurious rationality.

Thus, many times, the key to developing a competitive advantage is to just do it.

Strategies for Traditional Lenders

Unlike the rest of the agricultural industry where differentiation of product is increasing, the financing of agriculture is becoming more homogenous with money becoming a bulk commodity. Given these changes, it will be more difficult for traditional lenders to compete on price with some nontraditional lenders. In addition, with money becoming more of a commodity, production agriculture will begin to see credit tied less to a particular item than to the entire business (through an extended credit line).

While it may be difficult to compete directly on price, other avenues for competition still exist. These include partnering with other lenders, participation in a dual distribution system, and/or development of innovative products that may include equity and insurance products or some other niche-market focus.

Partnering With Others

Although it may be difficult for traditional lenders to compete directly with the credit card and point-of-sale financing strategies of input suppliers, some options are available. One alternative is to joint-venture with a major input supplier by placing bank or FCS lending personnel in the offices of the local dealer so that the farmer receives the benefits of "one-stop" buying and financing of inputs. A second alternative is to buy receivables from local input supply firms that extend operating credit to farmers and do so according to the standards and criteria set by the bank or FCS institution. A variation of this second approach is to provide a "floor-planning-like" financing package to the local input supplier that includes the traditional inventory and working capital financing and the funding efficiencies of point-of-sale origination and delivery, while enabling the traditional local lender to be an active participant in the market. Such joint arrangements would best include joint obligation for losses to reduce the incentive for credit extension to higher risk borrowers.

The development of specialized skills and techniques in data and information collection, credit evaluation, monitoring and collection, customer servicing, and sourcing funds suggests that, increasingly, the functions of marketing and customer relations, credit origination, collection and funding may be performed by different divisions of the same organization, or by different organizations integrated through alliances and joint ventures.

Computerized electronic credit evaluation and scoring techniques could reduce delivery costs while controlling risks. Likewise, use of specialized collection services—for example, telephone contacts combined with trained collection field staff—could be considered. These strategies would provide cost-effective delivery of additional credit services while controlling credit risks. Cost-effective, risk-controlled credit extension will be essential to maintain competitive interest rates for borrowers; increased competition in a slow-growth market will dictate that lenders be more rate- and service-competitive than in the past.

Dual Distribution System

While certain lenders may choose to compete on price, it may be possible to segment the market and compete on service. The existence of lenders that compete mainly on price and those that compete mainly on service would constitute a dual delivery system. Generic strategies focused on service include support and image differentiation. Delivery options for traditional agricultural lenders that allow differentiation based on service include regional loan production offices and direct lending using a "computers and cars" strategy. With portable computer technology and some customers' desire for personal service, the expectation that the borrower travel to the lender's relatively impersonal facility may be unrealistic. Instead, "going to the customer" with networking portable computers enables the lender to transact business in the borrower's place of business (or home) where he/she is more at ease, to access analytical software or data available only at the central offices, and to simultaneously complete an onsite inspection and evaluate the customer's operation.

Onsite contact may also allow the lender to evaluate the overall business rather than tying the credit to a specific asset. This total business line-of-credit would provide traditional lenders the opportunity to recapture some of the point-of-sale financing. The ability for complete credit services secured by the entire business could be an important mechanism to recapture lost market share from captive finance companies.

Niche Markets and Innovative Products

As consolidation of the financial sector proceeds, traditional lenders may want to focus lending activity on the market niche in which they have a competitive advantage. For example, many vertical coordination (contracting) arrangements drastically change the lending environment. Many of the contractual arrangements remove the need for traditional financing of production inputs with producers. This may cause a lender to shift from a line of credit to the financing of facilities as a chosen specialization. In addition, while inputs may no longer be financed by the small rural bank in the traditional manner, the integrator may still desire to finance these inputs via traditional channels. This may create an opportunity for other traditional lenders to capture this market, namely financing integrators.

Commercial banking and other financial services are merging between the commercial banking and the securities industries as regulatory changes and technological advances have begun to erode the boundaries created by the 1932 and 1933 Glass-Steagall Acts. Consolidation is also taking place in farming as family and commercial operations continue to grow in size and sophistication. Farms with increasing numbers of employees will require a wider range of financial products and services—including crops and personal insurance, retirement planning and accounts, payroll and direct deposit services, and agricultural product/input price risk management services. In addition, the demand for an equity financing product other than share leases will likely increase, given the increased risk borne by production agriculture under FAIR. The tax and legal implications of borrowing, leasing, estate planning, trusts, and changing accounting rules will require additional resources that agricultural lenders have typically not provided. Nor have they had to compete with local professionals who do provide these diverse services. There are limits to the services a lender can provide and continue to maintain quality service. However, customers' needs will be met by some organization. Lenders may need to arrange alliances to meet these financial needs.

Agricultural lending has traditionally been a relationship business where service was key. The definition of service is changing and one-stop shopping is coming to agricultural lending. Rural lenders will have to adjust to the increasingly complex needs of their customer base. Precision farming will require precision financing.

Conclusion

Traditional lenders in the future will look different than in the past. The information age will continue to lead to a consolidation in the number of lenders serving agriculture. Price competition will become increasingly intense. Those lenders that survive will do so based on a clear strategic focus upon which they excel. While money is becoming more of a "bulk" commodity, opportunities still exist for differentiation based on innovation, image, and support. Those lenders that do survive will have an accurate picture of the agricultural lending industry including, each competitor. The industry will not be comprised of the timid, nor will it be comprised of those for whom the status quo is acceptable.

References

Cole, C. A., A. M. Featherstone, and M. L. Albright. 1996. "Asset/Liability Management in Kansas Banks." Research Report #20, Department of Agricultural Economics, Kansas State University, Manhattan, Kansas, May.

Gabriel, S., and C. B. Baker. 1980. "Concepts of Business and Financial Risk," *American Journal of Agricultural Economics*, 62: 560-64.

Gilbert, Xavier, and Paul Strebel. 1991. "Developing Competitive Advantage," *The Strategy Process: Concepts, Contexts, Cases*. Henry Mintzberg and James B. Quinn (eds.). New York: Prentice Hall, pp. 82-93.

Gilder, G. 1981. *Wealth and Poverty*. New York: Basic Books.

LaDue, Eddy, and Marvin Duncan. 1996. "The Consolidation of Commercial Banks in Rural Markets," *American Journal of Agricultural Economics*, 78 (3): 718-20.

Mintzberg, Henry. 1991a. "Crafting Strategy." in *The Strategy Process: Concepts, Contexts, Cases*. Henry Mintzberg and James B. Quinn (eds.). New York: Prentice Hall, pp. 105-113.

Mintzberg, Henry. 1991b. "Generic Strategies." in *The Strategy Process: Concepts, Contexts, Cases*. Henry Mintzberg and James B. Quinn (eds.). New York: Prentice Hall, pp. 70-81.

Peters, T.J. "Strategy Follows Structure: Developing Distinctive Skills," *The Strategy Process: Concepts, Contexts, Cases*. Henry Minzberg and James B, Quinn (eds.). New York: Prentice Hall, 1991, pp. 809-14.

Porter, Michael E. 1980. *Competitive Strategy: Techniques for Analyzing Industries and Competitors*. New York: The Free Press.

PART THREE

WHO WILL BE THE LENDERS AND WHAT WILL THEY BE DOING?

6

The Role of Federal Credit Programs

Robert N. Collender and Steven R. Koenig

Federal intervention in agricultural credit markets is substantial and achieved through many mechanisms. Federal credit programs and government-sponsored enterprises (GSE's—see box) are the most visible types of federal intervention and are the focus of this chapter. However, federal supervision and regulation of financial intermediaries as well as monetary, fiscal, tax, and antitrust policies also substantially impact agricultural credit markets.

The federal government has been directly involved in agricultural credit policy since 1916 when Congress chartered the Federal Land Banks (FLB's), the first components of the Farm Credit System (FCS). Since then, an array of GSE's and federal agencies have been created to affect farm credit policy. Agriculturally oriented GSE's include the FCS (Federal Land Bank Associations, Production Credit Associations, Agricultural Credit Associations, Federal Land Credit Associations, the Federal Farm Credit Bank Funding Corporation, Farm Credit Banks, Agricultural Credit Banks, and Banks for Cooperatives) and the Federal Agricultural Mortgage Corporation (Farmer Mac). Federal agencies with roles in providing agricultural credit include the Farm Service Agency (FSA), the Small Business Administration (SBA), the Commodity Credit Corporation, the Farm Credit Administration, and the Farm Credit System Insurance Corporation. Other federal agencies or GSE's—including the Rural Electrification Administration, the Federal Home Loan Bank System, the Office of the Comptroller of the Currency, the Federal Deposit Insurance Corporation, and the Federal Reserve System—indirectly affect agricultural credit. In addition, many states have estab-

The views expressed are those of the authors and do not necessarily reflect the position of the Department of Agriculture or the current administration.

Box 6.1 What Is a Government-Sponsored Enterprise?

Since 1916 Congress has created a number of enterprises to improve credit availability and financial market competition to specific sectors of the economy including farming and rural areas, housing, and education. These government-sponsored enterprises (GSE's) include the Farm Credit System (FCS) and Federal Agricultural Mortgage Corporation (Farmer Mac) serving agriculture and rural areas; the Federal National Mortgage Association (Fannie Mae), Federal Home Loan Banks, and Federal Home Loan Mortgage Corporation (Freddie Mac) serving housing; and the Student Loan Marketing Association (Sallie Mae) and College Construction Loan Insurance Corporation (Connie Lee) serving higher education. Each GSE is privately owned and operated, limited to a specified economic sector and receives direct and indirect government benefits to help it accomplish its designated mission. GSE's hold over $1.5 trillion in financial obligations and have proven effective in promoting competition, lowering costs to eligible borrowers, and reducing regional differences in interest rates.

The two GSE's serving rural America are the FCS which has about $60 billion in loans outstanding and Farmer Mac which has about $500 million outstanding. The FCS lends directly to producers and harvesters of agricultural, aquatic and timber products, rural residents, agricultural cooperatives, farm-related businesses, and some rural utilities. It also has authority to finance the export of many U.S. agricultural products. The FCS can also lend to other financial institutions for short- or intermediate-term purposes if these institutions have a significant agricultural loan portfolio and a continuing need for nonlocal funds. Farmer Mac, which was established in 1988, guarantees the timely payment of principal and interest on loans originated and pooled by other lenders. Eligible loans include farm and rural home mortgages and certain FmHA-guaranteed loans.

---

lished their own agricultural credit programs and policies (Wallace, Erickson, and Mikesell 1994).

Federal support affects a large share of the agricultural credit market (Figure 6.1). At the beginning of 1997, the federal government directly held or guaranteed (directly or implicitly through a GSE) over one-third of all outstanding farm debt. In some regions, the federal government accounts for about half of all farm debt. Farm Credit System lenders hold a quarter of total farm debt, while USDA, through the Farm Service Agency, holds 6 percent directly and has guaranteed

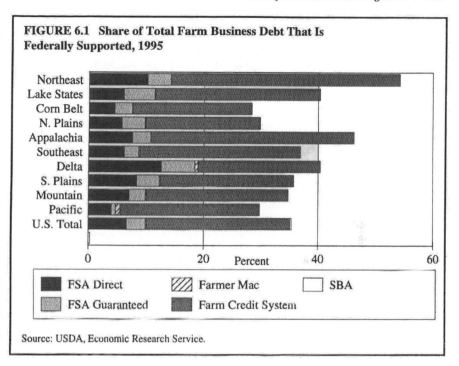

FIGURE 6.1 Share of Total Farm Business Debt That Is Federally Supported, 1995

Source: USDA, Economic Research Service.

another 4 percent of total farm debt (USDA 1997a). Farmer Mac, SBA, and other miscellaneous government sources of credit account for 1 percent or less each of total outstanding farm debt.

The many agencies and GSE's serving farm credit markets attest to agriculture's historic political, economic, and social importance. This array of farm credit programs and institutions has developed over the last eighty years in a piecemeal fashion. Over this period, the structure and technology of both the farm sector and of financial intermediation have changed dramatically. Such changes are expected to continue and, perhaps, to accelerate in the coming decades. Harrington and others (Chapter 2) indicate that agriculture has become more industrialized, market oriented, and dualistic (split between large industrial farms and small part-time farms) with a majority of food production arising from large commercial farms. Even small farms are no longer characterized by dependence on low, variable income from farming or by particularly low levels of income, wealth, and human capital (Hoppe et al. 1996).

Financial intermediation, once characterized by ubiquitous protection from competition and few consumer protections, now faces ever fewer barriers to competition, increasingly sophisticated abilities to process information, and considerable regulation to protect consumer interests. Changes in technology, law,

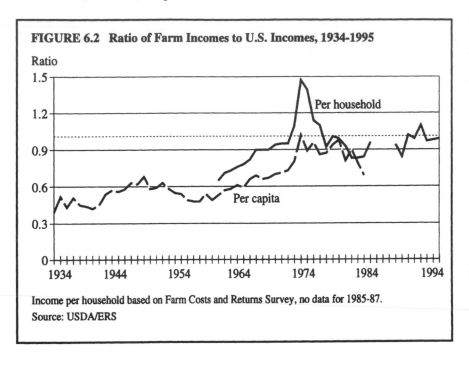

FIGURE 6.2   Ratio of Farm Incomes to U.S. Incomes, 1934-1995

Income per household based on Farm Costs and Returns Survey, no data for 1985-87.
Source: USDA/ERS

and regulation have all led to rapid consolidation within the finance industry, raising concerns among many bankers, policymakers, and rural residents.

Historically, federal intervention in agricultural markets, including agricultural credit markets, was closely linked to the "farm problem," which described prevailing conditions in agriculture for much of this century. The farm problem has been variously described to include "unfairly low" and highly variable incomes, a steep decline in farm numbers, and low rates of return on farm resources (Gardner 1992). At the turn of this century, "farm" and "rural" were essentially synonymous, so public policies that addressed the farm problem also addressed broader rural problems. Low farm-family incomes (Figure 6.2), difficulties in financing farm startups or the adoption of new technologies, the decades-long outmigration from rural farming communities (Figure 6.3), and perceptions that private lenders were not adequately, efficiently, or fairly supplying credit to farm owners and operators have all motivated federal intervention in farm credit markets. Rent-seeking and other political considerations have also motivated federal intervention. While these are acknowledged to be important practical reasons for the policies that are in place, this chapter focuses on the economic rather than political merits of federal intervention.

Politics and rent-seeking aside, motivations for federal intervention in farm credit markets can be summarized into two categories: enhancing economic efficiency and addressing concerns about social equity. Policy initiatives designed to

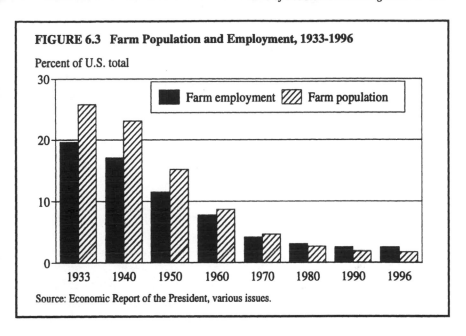

FIGURE 6.3   Farm Population and Employment, 1933-1996

Percent of U.S. total

Source: Economic Report of the President, various issues.

strengthen credit market efficiency can increase the resources available to society and therefore, are potentially self-financing. Credit policies that increase lender competition, lower transaction costs, or improve information availability enhance market efficiency and hence facilitate a more productive allocation of resources.

Policies and programs that advance social goals through credit markets typically create economic inefficiencies and require ongoing government expenditures. Such credit policies may seek to channel loanable funds to promote such policy objectives as preserving existing businesses, equalizing economic opportunities, or reducing income disparities. Since such programs seldom directly address underlying causes of social inequality, they tend to be inefficient at achieving their ends.

This chapter proceeds as follows. First, we describe some of the past federal actions to address market efficiency and social equity concerns, their rationales, and available evidence regarding their effectiveness. Next, we present some of the economic arguments for and against continued intervention in these markets. We close the chapter with an assessment of the federal role in these markets in the next few decades given the evolution of both U.S. agriculture and U.S. financial markets.

A Brief History of Federal Intervention
in Agricultural Credit Markets

Relieving the farm problem and achieving other social objectives have arguably been the primary motivations for federal intervention in farm credit markets. Federal credit policies have long sought to directly influence the distributions of income and wealth. Concern about inadequate rural incomes has prompted Congress to make loans available to borrowers who were not creditworthy (Brake 1974). These credit policies have been carried out principally through the FSA and its predecessors (Farmers Home Administration, Farm Security Administration, and Resettlement Administration).

Direct federal intervention in farm credit markets dates back to 1918 when President Wilson set aside $5 million in defense and national security appropriations to provide low-interest loans in drought-stricken areas. Initiated near the end of World War I, these loans were intended to maintain farm production, improve farm income, and keep farmers on the farm. Federal emergency lending continued sporadically until 1931, when federal farm emergency loans were made nationally available on a permanent basis. While subsidized credit programs were initially a targeted response to natural disasters, programs were later based on either general economic conditions or the specific economic circumstances of eligible borrowers.

Much of the current federal intervention in farm credit markets can be traced to policies that arose during the Great Depression in the context of the New Deal. Subsidized credit became entrenched in federal farm and rural development policy, and the FCS was reconfigured, recapitalized, and enlarged. Social stability and political support were major motivations for the increased role of federal credit—important considerations given that 30 percent of the total U.S. population resided on farms.

Government intervention in farm and rural credit markets was accomplished largely through supervised low-cost loans and grants. The Resettlement Administration, and later USDA's Farm Security Administration, made low-cost loans to assist poor or tenant farmers in purchasing farms under the presumption that owners were better stewards of land and that widespread land ownership would improve rural incomes and wealth. Low-cost loans were also intended to help farm families remain on their farms or reestablish themselves in farming. These programs also improved economic efficiency by providing liquidity to farm capital markets that suffered a withdrawal of many suppliers of debt capital (banks, life insurance companies, and individuals).

The FCS was an integral part of federal credit policy and was given responsibility to administer some federal farm lending programs. As with direct loan programs, FCS lending was intended both to address market failures and to achieve social goals, but with fewer direct subsidies. Production Credit Associations were added to provide a more stable source of short-term financing, primarily for small and family-sized farmers. FLB loan rates were written down by the U.S. Treasury, and Land Bank Commissioner loans made by FLB's were used to refinance existing farm indebtedness, including many nonperforming FLB loans.

After World War II, Congress changed the orientation of farm credit programs and consolidated them into a newly created Farmers Home Administration (FmHA). With tighter labor markets, the emphasis on maintaining farm populations diminished. The program focus shifted to supervised operating and purchase loans to small and family-sized farms unable to qualify for commercial credit. Federal credit programs also provided a broader safety net through emergency lending. Throughout the 1960s and 1970s, eligibility criteria and the scope of lending activity were repeatedly expanded, and, by the 1980s, FmHA was a major provider of farm capital, holding or backing over 17 percent of total U.S. farm debt.

The FCS was largely relieved of any Depression-era social objectives after World War II, and the FCS continued to lend mainly to commercial-sized farms (Brake 1974; Koenig and Dodson 1995). Legislation and regulation have expanded the scope of its authority, while eliminating most targeting requirements.

In contrast to the direct federal lending programs, other federal programs—including the FCS and Farmer Mac—were largely created to address real or perceived credit market imperfections. The FCS addressed imperfections stemming both from the geographic isolation of many farming communities and from banking laws and regulations that limited the activities and geographic span of banks and thrifts. Interstate banking was prohibited and branching was restricted in most states, forcing banks and thrifts to rely heavily on local deposits. Because local deposits were banks' and thrifts' primary source of funds, credit availability and pricing varied depending on the balance of deposits and local lending needs. In addition, restrictions on branching and chartering of banks restricted competition in some local banking markets.

Agricultural lending was often not attractive to commercial banks compared with other uses of loanable funds that were thought to be more profitable, easier to administer, and safer. Banks' reliance on short-term deposits, which were payable on demand, made longer-term farm mortgage loans too risky. Agricultural income was more volatile than income of many other commercial businesses, making repayment more uncertain.

The Small Business Administration (SBA), another example of federal credit market intervention justified on the basis of economic efficiency, was predicated on the failure of commercial banks to offer medium-term loans, including loans to agriculture. Continuation of the SBA is partly based on the premise that banks are risk-averse and, therefore, deny some loans that could be expected to have positive economic returns. This divergence of private and public interests reduces economic efficiency (Rhyne 1988). Similar arguments could be made with respect to USDA's farm loan programs.

The Federal Agricultural Mortgage Association (Farmer Mac) was also created to increase market efficiency by operating a secondary market primarily for farm real estate loans. Commercial banks hoped an active secondary market for farm real estate loans would provide both additional liquidity and a reliable, low-cost source of long-term, fixed-rate funding to improve the banks' ability to compete with other real estate lenders, including the FCS and insurance companies. To date, Farmer Mac has failed to develop as hoped and has little impact on bank competitiveness or overall market efficiency.

Past Credit Program Effectiveness

Federal credit programs have had many specific objectives pertaining to market efficiency and social equity. Such objectives include raising farm income, reducing rural-to-urban migration, enhancing credit market competition, decreasing interregional variability in borrowing costs, providing liquidity to rural capital markets, and reducing credit rationing. Below, we review available evidence concerning the effectiveness of federal action in achieving each of these specific objectives.

Income Support. Policymakers often resort to subsidized agricultural credit programs to reduce disparities between rural and urban incomes or wealth. Some subsidized federal credit programs have been based on the premise that access to capital was the primary constraint preventing farmers (or other targeted groups) from achieving adequate income or efficient scale, while others have provided credit to replace lost income from natural disasters or economic emergencies (Brake 1974). In general, both types of credit programs have been largely ineffective at achieving their goals and have had unintended consequences.

Four studies have significant implications for understanding the impact of USDA's lending programs. One study by LeBlanc and Hrubovcak (1986) measured the effects of tax policy on agricultural investment. Their model included borrowing costs, and their results were extended by Bosworth, Carron, and Rhyne (1987), who concluded that farm loan subsidies are primarily income transfers, which are captured by land owners rather than by borrowers. This occurs because the supply of farmland is relatively fixed compared with other factors of farm production. Another study, by Hughes, Penson, and Bednarz (1984), found that government farm credit policies have a substantial impact on farmland prices. They concluded that subsidies increase land values, farmers' holdings of financial assets, and farm debt, but find little long-term impact on total amount of farm real estate in production or the distribution of farm real estate ownership. Gale (1991) showed through simulations that government credit programs increased funds to targeted sectors, while reducing funds to non-targeted sectors. Under the most favorable assumptions, Gale concluded that total private borrowings increased slightly (about 5 percent), but the cost per dollar of added investment—more than 40 cents—was high, and welfare losses to the economy were large.

These studies indicate that the unintended consequences of subsidized credit programs include increases in rents or prices that benefit landowners, not the poor; concentration of production (see below); and concentration of benefits determined by loan size—larger loans mean larger benefits. In particular, subsidized federal emergency lending programs have been very costly. Often, these programs have been poorly targeted, and their loan delinquency rates dwarfed those of other USDA farm lending programs, hovering above 30 percent for over a decade. Of the over $29 billion in emergency lending for fiscal 1975 through fiscal 1996, over $12 billion was written off. This experience has several possible explanations. For example, replacing lost income with credit is unlikely to work if loan proceeds are used to support consumption rather than investment. Such credit-based consumption leaves the borrower a debt burden without a source of additional income from which to make payments.

Structural Adjustment. Federal intervention in agricultural credit markets has sought to influence the structure of agriculture and to slow outward rural migration. However, this policy is inherently self-defeating. Subsidized credit has been used to help inefficiently small farms increase their size, thus encouraging consolidation of productive resources. Given the relatively slow historical growth in farm product demand and quality farmland, not all farms can grow and remain profitably in business.

Following a temporary runup in farm commodity prices in the early 1970s, USDA rapidly enlarged its (subsidized) farm lending programs, lending $35 billion during the eight fiscal years beginning in 1974 through its Farmers Home Administration loan programs alone. USDA credit was often heavily subsidized and arguably facilitated investment to expand productive capacity at a time when Earl Butz, the Secretary of Agriculture, was exhorting farmers "to plant fence row to fence row" to feed a growing world population. This subsidy, along with other farmer-friendly policies and subsidies, may well have exacerbated the supply response that ultimately burst the agricultural sector asset bubble in the 1980's. Much of this government credit, especially that extended through expanded emergency loan programs, was not targeted to smaller or limited-resource farmers because eligibility requirements were weak or not strictly enforced. Therefore, some credit enabled the rapid expansion of already large, often creditworthy farms.

The behavior of the FCS may also have amplified the boom and bust cycle of the 1970's and 1980's. The Farm Credit Act of 1971 liberalized and broadened the lending authorities of the FCS, probably increasing investment rates during the following boom years (Carey 1990). After passage of the 1971 Act, FCS lending volume soared as the System used its liberalized lending authorities and GSE funding abilities to grow during a period of rising interest rates. By pricing its lending activities at average cost instead of marginal cost, the FCS underpriced capital costs and hence stimulated investment and consolidation more than might have occurred.

Using subsidized credit can no longer be justified on the basis that labor markets cannot readily adjust to flows of labor between farming and other sectors. Regardless of past polices or economic factors, farm production is now concentrated and increasingly like other industries. The net effect of federal credit policies on farm numbers and structure is uncertain and depends on targeting and other policy objectives. In the 1990s, federal farm credit programs primarily target the least creditworthy farm operations, as they largely did prior to the 1970s. Subsidized credit often boosts cashflow sufficiently for uncreditworthy (unprofitable, inefficient) borrowers to remain in farming longer than they otherwise could. Thus, credit programs have sometimes slowed the migration out of agriculture. But, subsidized credit has also helped some farms achieve an economically viable size, hastening consolidation of farm production.

Again, subsidized credit policies tend to boost asset values, benefiting owners of capital assets more than targeted populations. Land values increase because subsidized interest rates increase the prices buyers are willing and able to pay. These benefits come at the expense of taxpayers, borrowers, and competitors. Federal credit policies may also contribute to the consolidation of farming assets in other unintended ways. For example, both the FCS and Farmer Mac focus their lending on larger commercial farm operations (Brake 1974; Dodson and Koenig 1994; Koenig and Dodson 1995; Johnson 1963; Koenig and Ryan 1997), perhaps giving these producers an additional competitive advantage relative to small and medium-sized operations.

Overcoming Discrimination. Correcting credit market failure due to discrimination is another objective of federal credit policies. Federal farm credit programs have long sought to improve credit availability to minority and limited-resource farmers, and, in recent years, funds have been targeted to serve these groups. But, the minority farm population continues to decline (Kalbacker and Rhoades 1993). Minority farm numbers dropped from 54,000 in 1982 to 43,000 in 1992, or 2.3 percent of all farms. While no quantitative evidence is available on the effectiveness of federal credit programs in assisting minority farmers, a recent USDA report (1997b) documents minority farmer complaints that USDA has done more to hurt than to help small and minority farmers. Such complaints include unfair lending practices, systematic exclusion, racial and gender bias, and neglect. In addition, rural GSE's have no effective targeting requirement in this area and likely lend very little to these groups, although statistics are not readily available (USDA 1997c, p. 31).

Stimulating Production and Infrastructure Development. Federal credit programs have also sought to encourage capital investment to stimulate production or to encourage targeted investments in infrastructure. One justification for an expanded role of federal credit programs in the 1970s was to ensure that ample capital was available to expand production to meet perceived shortages of agricultural commodities. USDA's grain storage and drying facility loans in the late 1970s and early 1980s sought to encourage capital investment in farmer-owned grain storage capacity. This infrastructure financing was part of an attempt by the government to raise and stabilize farm prices and incomes by reducing the flow of grain to the market during periods of excess production. While storage facility loans are no longer part of this policy, the government continues to provide subsidized commodity inventory loans for some commodities in an attempt to support and stabilize farm prices and incomes.

Subsidized credit has been instituted to encourage investment in projects with certain other social objectives, including natural resource development and natural resource conservation. Credit policies designed to encourage particular investments—for example, to improve water quality or reduce soil erosion—have generally been replaced with cost-sharing grants because of lower delivery costs. LeBlanc and Hrubovcak (1986) found that loan subsidies are inefficient for promoting agricultural investment because reductions in interest rates increase demand for land by twice the increase in the demand for equipment and structures. Given the relatively fixed supply of agricultural land, the subsidies largely benefit landowners rather than borrowers.

Enhancing Retail-Level Credit Market Competition. The FCS was organized as a network of retail lending cooperatives because of perceptions that rural financial markets were not competitive. In theory, FCS's mandate to serve agricultural producers nationwide enhances competition in local credit markets. Although the vast majority of rural credit markets' banking activities are highly concentrated and the associated efficiency costs from this concentration may be sizable, it is less clear that FCS activity has significantly improved market performance. This uncertainty stems from past evidence and several characteristics of the FCS. In a comprehensive evaluation of the historical impacts of FCS activity on farm credit markets, Johnson (1963) found a decline in the average differential between farm mortgage rates and rates on high-grade bonds for all lenders. He attributed the decline to the Land Banks and insurance companies that made loans at smaller differentials than did commercial banks, but did not find a sustained decrease in the differential for commercial banks that would indicate the loss of market power.

Current law and regulations allocate exclusive territories to most FCS lenders. Since not all FCS lenders actively use all FCS lending authorities, in most markets FCS authority introduces a maximum of one additional competitor. That competitor may, at times, be limited in its short-run ability to respond to increases in economically viable demand for loans. These characteristics severely limit the potential impact of FCS activity on market concentration or performance. In contrast to the FCS, Farmer Mac can encourage new entrants into local markets because access to its funding is not limited by lender type. However, Farmer Mac provides a much narrower funding mechanism and has yet to prove its effectiveness.

In addition, the structure of direct federal and state programs, GSE charters, and banking laws has encouraged division of agricultural loan markets into three segments: (1) struggling and low-resource farms are served mostly through federal and state direct and guaranteed loan programs; (2) lending to part-time farmers is dominated by commercial banks, and (3) bigger, full-scale, commercial-size farm lending is dominated by FCS lenders and insurance companies. (See Koenig and Dodson 1995, for further details on market segmentation.) Various barriers and competitive advantages make this segmentation sustainable, including subsidies, capitalization rules, local physical presence of lenders, and organizational structures. Such segmentation is also evidence that competition among lenders remains imperfect despite the presence of the retail-level GSE.

Reducing Variation in the (Risk-Adjusted) Cost of Credit Across Regions. Some evidence exists that federal credit programs have helped equalize the cost and availability of credit in local markets. Both USDA's farm credit programs and the FCS have mandates and operations that are national in scope and offer credit to farmers on fairly uniform terms nationwide. Johnson (1963) documented that interregional differences in farm mortgage rates fell markedly in the decades after the Federal Land Banks were established. However, he also documented similar changes in interregional interest rate differentials for other types of loans. He concluded that the "Land Banks probably had the effect of hastening the reduction of regional differentials in farm mortgage interest rates, but may not now have much effect on such differentials (p. 281)."

Providing Liquidity to Rural Credit Markets. During both the Depression and the 1980s agricultural financial crisis, government credit programs provided considerable liquidity to stressed agricultural credit markets. In 1933, the FCS was recapitalized and restructured to help all farmers, especially those in financial stress. The government provided credit assistance directly through the FCS during this period, not just to assist existing FCS borrowers. Interest rates for FLB loans were frozen at 4.5 percent with the Treasury paying any difference. The Treasury paid in capital for loan extensions and five-year principal deferments. During the 1980s crisis, the FmHA doubled its annual total farm lending from $3 billion in 1983 to $6 billion in 1986 and provided special assistance and loan guarantees to farmers and banks facing financial distress.

Unfortunately, no research has been published addressing the costs and benefits of these actions. It is possible, given the relatively short and severe nature of these periods of distress, that the net benefit of providing additional liquidity could be greater than at other times. However, it is not clear that the net benefit is positive. Johnson (1963) argued that funds provided through the FCS probably crowd out other sources, especially given that fairly conservative lending practices prevailed within the FCS prior to the 1970's. Similar lending practices have prevailed among FCS lenders since legislative reforms passed in 1985 and 1987. Carey (1990) has argued that the massive increases in FCS (and FmHA) lending in the 1970s were destabilizing to the farm economy. However, Stam

(1995) finds no consensus among researchers on the effect of credit availability on the value of farmland, the primary capital asset in agriculture.

Reducing Credit Rationing and Redlining. Credit rationing is usually defined as the situation where, among identical borrowers, some are financed and others are not. Redlining refers to a situation where a class of (creditworthy) borrowers is refused credit entirely. Both credit rationing and redlining represent market failures and, therefore, areas where government action could improve economic performance. Using data from 1977-1984, Calomiris, Hubbard, and Stock (1986) found evidence of credit rationing to agriculture. USDA's farm credit programs, which are intended to serve as agricultural lenders of last resort, and FCS's mandate to be "responsive to the credit needs of all types of agricultural producers having a basis for credit" are consistent with addressing these market failures. Again, while it is possible that government-supported lending programs have, especially at times of widespread distress, addressed such failures, they have probably not usually done so in a cost-effective manner, given the results of Gale (1991) and Rhyne (1988). The probability that these programs have been cost-effective is reduced by the fact that they have received substantial explicit and implicit subsidies.

Changes in Agriculture and Policy Have Made Subsidized Credit Programs Outdated

It is increasingly difficult to make a case for subsidizing credit to agriculture. Using subsidized credit can no longer be justified, as it was in the past, on the basis that labor markets cannot readily adjust to flows of labor between farming and other sectors. As noted earlier, farm production is now concentrated and increasingly like other industries. Providing subsidized credit of any kind to the most productive farms is difficult to justify from an equity or an income perspective. Just 500,000 of the nation's roughly 2 million farms account for nearly 90 percent of total annual farm sales, and just 100,000 farms account for nearly 40 percent of the total. Dodson and Koenig (1995b) found these 100,000 very large farms owed 25 percent of total farm debt, reported average net farm incomes approaching $100,000, and mostly had networths over $1 million. These firms demand more and larger loans (indebted farms average nearly $400,000 in debt) and hence can negotiate for debt from a wider range of sources. They are better able to shop nonlocal credit markets, such as regional and national lenders (e.g., life insurance companies) because of the size of their credit needs and the range of financial services they require. These borrowers are more attractive to lenders because fixed lending costs can be spread over larger loan balances and multiple services. Industrialized and vertically integrated farms are more likely to require more complex financing and financial services, such as cash management accounts and financial leases, that are more likely to be provided by regional and national lenders.

As agriculture industrializes and vertically integrates, farm operators are shifting to other forms of financing, including production contracts. Dodson (1997) found that almost 30 percent of all farm production is produced under some form of production contract. Growth in production contracting, especially in the broiler and pork industries, has reduced financial risk to producers as the price and/or production risks are shared with or shifted to integrators. In these and other sectors of the farm economy, integrators frequently offer credit to contracting farm operators as well.

Small farms represent another group where subsidized credit may also be difficult to justify from either a social-equity or an income-support perspective. Half of all U.S. farms produce less than $10,000 in sales and three-quarters of farms have sales under $50,000. These 1.5 million noncommercial farms account for about 13 percent of total annual farm sales. But, because the majority of these farms receive most of their incomes from off-farm sources, farm income is a relatively minor component of their total household income. Many of these farms exist because their operators value a farm lifestyle or are semiretired farmers. About 400,000 small farms exist with low off-farm incomes where subsidized credit may have a greater impact on total household income. However, the direct impact of credit subsidies is limited by the fact that less than a third of these farms carry any debt, and for those that do, total outstanding debt averages just $50,000. Furthermore, 250,000 of these farms have owners over 60 years of age who are unlikely to have the interest in or ability to expand their farming activities.

Subsidized credit has also been used to provide disaster relief, but subsidized crop insurance has largely replaced *ad hoc* disaster relief and credit programs. Price and production risks associated with the variability in weather patterns and production cycles have been primary arguments for government intervention in U.S. agriculture. Congress since 1918 has provided *ad hoc* disaster aid programs to shore up farm income and wealth following natural disasters, such as floods or droughts, or periods of low farm income or prices. Credit has been a prominent feature of these programs. From fiscal 1980 to 1984, 60 percent of the nearly $26 billion in FSA's total lending was for disaster-related lending programs, both for economic and natural disaster purposes. The 1985 farm bill made changes to USDA's emergency disaster loan program to reduce the cost of *ad hoc* disaster relief. Such changes included requiring loans to go only to family-sized farms, limiting loan size to $500,000 or less per disaster to those who have been denied credit from other sources; spending guidelines were sharply lowered, as well.

Subsequent farm bills have further restricted eligibility and use of disaster loan programs. Between fiscal 1990 and 1994, just 15 percent of the $12 billion in total FSA lending was for emergency disaster purposes. Also in the 1985 legislation was a requirement that production losses that could have been covered under the Federal Crop Insurance Act were no longer eligible for disaster loans. This policy has been extended to eligibility for *ad hoc* disaster relief. The Federal

Crop Insurance Reform Act of 1994, as amended, strengthened reliance on subsidized insurance to provide a safety net for producers. Subsidized crop insurance is available for a wide range of crops across the country and participation in crop insurance programs is high.

Passage and implementation of the Federal Agriculture Improvement and Reform Act of 1996 could generate greater political pressure to use farm credit programs to maintain family-sized farms and to help farmers when incomes are low. As discussed in chapter 2, this act was an historic step toward removing the control over the supply and price of major agricultural commodities that has been in place since the 1930s. Removal of supply management could increase variability in income levels and credit demands for some farmers (Glauber and Miranda 1996), and average income levels could fall if the transition payments provided through 2002 are not replaced with some other form of support. Deregulation of farm production could contribute to further consolidation of farm productive assets. Without commodity programs, many producers have less of an income safety net, and operators with good production management but poor marketing practices will be more likely to exit. However, the transition to a market-oriented agricultural economy will not make credit programs more effective than they have been in the past. Other policy instruments, including lump sum payments and well-managed insurance programs, will be more effective at achieving most policy goals.

Economics of Credit Market Intervention

The two broad justifications for intervention in financial markets—to enhance market efficiency or to further social equity—have very different economic implications. If successful, the former directly enlarges the economic pie, while the latter usually involves rearranging the sizes of the servings and may well cause the total pie to be smaller. That is, programs to attain "social equity" through credit markets are almost always inherently inefficient, creating economic distortions, lowering economic growth, and requiring government expenditures. Such programs are more efficient when they are an integral part of programs that educate or train recipients to succeed without the need for preferential financing, and when they facilitate rather than obstruct economic adjustment.

Importance of Economic Efficiency

Research indicates that largely private, competitive financial systems are closely associated with sustained, above-average rates of economic growth (King and Levine 1993). Financial markets affect economic growth rates through the role of private lenders in allocating capital to businesses. Private lenders must choose among borrowers with productive uses for capital and, in the process, allocate capital among competing uses. If private lenders make loans that do not generate sufficient profits to allow repayment with interest, capital has been poorly allocated and economic growth rates suffer. Public lenders, because of political pressures, usually fail to allocate capital efficiently and often do not intend to do so.

In a perfectly competitive financial system, credit market inefficiencies could not exist for long. However, legal barriers to competition and the cost of obtaining information make market segmentation possible, reducing competition. Federal policies that heighten lender competition, lower transaction costs, or improve information have enhanced efficiency in such financial markets. Successful "market efficiency" programs are self-sustaining and do not require continued government subsidies. Sometimes, subsidies accompany these programs to enhance their initial acceptance and ability to compete against entrenched, noncompetitive institutions.

Some market efficiency programs, including the FCS and Farmer Mac, were chartered to overcome barriers to competition or financial flows caused by restrictive banking laws and regulations, asymmetric information, geographic isolation, and limited communications technology. In many cases, financial markets in the United States were purposefully segmented. Until recently, many states restricted within-state and interstate branching by financial institutions. The Glass-Steagall Act segments commercial banking markets from investment banking and insurance markets. And many other laws and regulations reduce the ability of financial institutions to channel funds efficiently to their most productive uses.

Some Economic Issues Regarding Social Equity Programs

An efficient, competitive financial market offers buyers of credit services equal opportunities, but even this equality of opportunity may not yield a "socially equitable" allocation of resources by various noneconomic measures. For example, the existing uneven distribution of education and wealth within the U.S. population creates an uneven distribution of creditworthiness that may be politically unacceptable. Thus, many government programs exist because of dissatisfaction with the way private, competitive markets allocate capital resources. Concerns over "fair" treatment of particular populations (e.g., low-income households and family farmers) underpin some federal credit programs.

Market interventions that alter the allocation of capital for noneconomic reasons inevitably lead to economic losses and slower economic growth. Such interventions include targeted, subsidized, or allocated credit, or artificially low interest rate ceilings on loans.

Some social equity concerns, such as nondiscrimination, are consistent with market efficiency. Most, however, are not related to efficiency and can not be effectively addressed through credit programs because of the nature of money. Credit programs are poor vehicles for transmitting subsidies because loan funds may be used for unintended purposes (the fungibility problem), the borrowers may have access to credit from other sources, the subsidy benefits may accrue to private lenders rather than to targeted borrowers, or favorable terms of credit may be capitalized into the values of the assets being financed (Barry and Associates 1995). Fungibility has been a problem for government lending programs that try to increase investment in particular sectors. Qualifying borrowers often use subsidized credit for nontargeted purposes, while much of the targeted investment might have been funded by other, unsubsidized sources. Improved or subsidized credit availability potentially lowers the cost for any use of capital, not just the use intended. If the relative attractiveness of various investments is otherwise unaltered, preferential credit for one sector may fund more attractive profit opportunities in other sectors.

Governments as Well as Markets Are Imperfect

Government failures, enforcement costs, and the nature of loan contract obligations also mitigate the effectiveness of credit in attaining social equity goals. Government programs suffer from their own set of imperfections, and government remedies do not axiomatically improve social outcomes compared with the outcomes of imperfect markets. Society must choose between imperfect market outcomes and imperfect government remedies.

The case for government intervention is strongest where markets are most imperfect. For example, in the areas of national defense, police protection, education, roads, sewers, public health services, and environmental degradation, persuasive externality and public goods arguments can be more readily made. Interventions designed to encourage or discourage private sector activity (e.g., tariffs, quotas, excise taxes, subsidies, licenses, price controls, limitations on entry, and mandates of various kinds) or direct government ownership of facilities that are capable of private sector operation create more intractable problems. Such interventions can improve some aspects of market outcomes, but often seriously diminish growth because of distortions that accompany government action. Distortions arise both from providing improperly priced goods and services and from offsetting attempts at revenue collection.

In general, government programs face a pervasive set of problems that diminishes their ability to effectively address social or efficiency concerns through direct economic action (White, 1994). These problems include difficulty in formulating clear implementable goals, establishing appropriate incentives for both program staff and targeted beneficiaries, and minimizing rent-seeking or rent-capturing behavior. Each of these problems has occurred with respect to both direct federal agricultural credit programs and GSE's serving agriculture.

Benefits and Costs From Federal Credit Market Interventions

Recent studies (Duncan, 1997; RUPRI, 1996; U.S. General Accounting Office, 1997; USDA, 1997c) of rural credit markets have concluded that imperfections continue to exist in rural credit markets, leaving the door open for a government role. The evidence is much less compelling for commercial agriculture, but many farms, especially smaller ones, face the same limited rural capital markets as do other rural households. This conclusion, that markets are imperfect, raises the issue of what the benefits and costs might be to improving them. A related set of issues involves the benefits and costs of using financial markets to address social concerns.

In evaluating the relative magnitudes of the costs and benefits of possible federal credit intervention, important factors include

- the economic losses from market imperfections or social "injustices" that such activity is likely to address,
- the extent to which lending currently provided on less subsidized or unsubsidized terms would shift to new programs and the extent to which additional funding would finance economic activity that would not otherwise occur, and
- the extent to which any additional activity would otherwise have been undertaken without additional federal benefits.

Improving Market Performance. Benefits can arise from such market-performance improving outcomes as reducing market power, lowering transactions costs, or spurring investments in research or technology that shift the marginal cost curve for the financial services industry to the right. While the vast majority of rural banking markets are highly concentrated and the associated efficiency costs from this concentration may be sizable, it is unclear that additional federal activity would significantly improve market performance. For example, Berger and Hannan (1994) indicate that market concentration must be reduced to a level equivalent to ten equally sized competitors for net benefits to accrue. In addition, most proposals that ostensibly address rural market shortcomings have a large potential to provide unearned economic rents and are promoted by interest groups who would capture those rents.

The predominant avenue for federal lending aimed primarily at improving credit market efficiency has been the GSE's. Yet, several characteristics or practices of rural GSE's reduce the likelihood that they are both willing to and capable of improving market performance with respect to many agricultural borrowers:

- Farm GSE's have tended to serve the least risky and most profitable market segments (Dodson and Koenig, 1995; Collender and Erickson, 1996),

- FCS lenders are generally granted exclusive territories, limiting their impact on market concentration, preventing them from actively competing with each other, and preventing successful or innovative associations from serving borrowers outside their territories,
- FCS capitalization practices and regulations tend to limit their abilility to compete in rapidly growing or changing markets,
- Farmer Mac has demonstrated little ability to penetrate many agricultural real estate markets, and
- advances from Federal Home Loan Banks (FHLB's) do not diminish market power of retail competitors by encouraging new entry.

For benefits to accrue, federal lending activities must be financed that currently are not. If lending shifts from private sources to federal or federally sponsored sources, even if at a lower price, no net gains accrue to the economy—borrowers gain at the expense of taxpayers and competitors. This crowding out provides no economic benefits on its own. If financial services are provided to previously unserved or underserved borrowers or other lenders operate more efficiently, then economic benefits may accrue to the economy.

The degree to which subsidized activity serves new borrowers or existing borrowers is important in determining its economic impacts. Gale's study (1991) of the efficiency effects of federal credit programs found such programs are seldom cost effective despite the fact that target groups often experience large gains in borrowings/credit use. Cost effectiveness is hampered by the fact that these programs are unable to exclude groups that would otherwise be served and because no societal benefits accrue from subsidizing such groups. Thus, on a national scale, the characteristics of federal credit interventions make it unlikely that expanded activity at either the retail or wholesale level would substantially improve agricultural credit market structure or performance.

Social Benefits of Federal Activity. Congress often charges both federal direct and guaranteed lending programs and GSE's with pursuing social goals beyond enhancing market efficiency. GSE's are often required to provide benefits to specific groups, not just the most profitable market segments within their granted authority. Federal programs are often directed to provide benefits to specific groups, such as minorities or beginning farmers. Meeting social equity objectives is encouraged by targeting a portion of a GSE's business activity to segments of the population—typically defined by income class or geography—thought to be inadequately served by private sector credit markets.

Many rural areas could benefit from additional economic opportunities based on per capita income and earnings per job. For example, incomes are relatively low and poverty is widespread in the 535 counties that ERS categorizes as being "persistent poverty" areas—those where at least 20 percent of the population were below the poverty level in each of the years 1960, 1970, 1980, and 1990 (USDA 1997c). However, no evidence exists that credit availability, specifically for agriculture or in general, in these areas is a primary barrier to improved economic performance. In addition, little evidence exists of widespread farm credit problems since the mid-1980s. Even for that period, agricultural liquidity and credit availability problems were concentrated among about 10 percent of farmers who held about a third of all farm debt (Stam, Koenig, Bentley, and Gale 1991).

Measuring the social benefits of increased credit activity is difficult. Social benefits accrue when targeted groups gain access to financing that would otherwise be unavailable to them, and when they are able to use this access constructively. Often nonfinancial factors such as knowledge, skills, abilities, and access to markets for inputs or for production determine whether benefits accrue. If these and other factors are important to determining wealth and income of targeted groups, then merely expanding access to credit is unlikely to alleviate social concerns or meet specific social goals.

Targeted credit can raise economic costs if the targeting carries a substantial degree of subsidy and does not counter market imperfections. Subsidies distort the financial incentives of the targeted population and reduce the efficient allocation of credit by lenders. Nevertheless, targeting can bestow economic benefits to groups not served competitively by private credit markets because of noneconomic factors, such as location or discrimination. To be cost effective, social benefits need to be targeted toward those who are not served by commercial lenders and can benefit from additional credit. Credit provided to support consumption is inefficient relative to direct payments and undermines the obligation of repayment inherent in the concept of credit.

GSE's serving both agriculture and housing have a mixed record of achieving social goals. Groups targeted for federal lending often experience large gains in credit use, but because programs affect both target groups and groups that would otherwise be served, such programs are not always cost effective (Gale 1991). Yet, direct federal lending programs may be more successful at targeting designated groups than are GSE's (U.S. Department of the Treasury 1996; Dodson 1996a).

Rhyne (1988) points out in her study of SBA lending that loan subsidies do not increase total funds in capital markets, so interest rate subsidies to one group result in higher rates for other borrowers. Thus, costs may substantially exceed the obvious direct costs to taxpayers. Unless social benefits surpass direct costs by a large amount, programs reduce efficiency of resource use. Her conclusion, consistent with many others (for example, Gale 1991; Buttari 1995; Adams, Graham, and von Pischke 1983), is that subsidies are an inefficient means to allocate resources or to redistribute income. Programs usually cannot be shown to achieve aims or to generate benefits sufficient to justify costs, often resulting in transfers to politically powerful groups.

Future Directions for Government Intervention
in Agricultural Credit Markets

The persuasiveness of rationales for federal intervention in agricultural credit markets—enhancing market efficiency and meeting social objectives—has diminished over the past eighty years. Yet, the federal role in agricultural markets remains high, with over a third of farm debt either directly or indirectly guaranteed by the federal government. A recent study (USDA 1997c) concluded that credit markets serve agriculture better than other rural sectors and that little evidence exists of widespread rural credit market inefficiencies or market failures.

The future structure of agricultural production (chapter 2) and financial markets will continue to shape agricultural credit demand, supply, and the federal government's role. Agricultural production has experienced rapid concentration during the twentieth century, and little reason exists to expect this trend to slow or reverse. Evolution in farm structure has been and will continue to be driven by technological innovation, but agricultural and economic policies will continue to affect the speed of adjustment. Vertical integration and horizontal consolidation will continue as producers adopt new or existing technologies and exploit readily available economies of size, scale, and scope.

Over the last eighty years, agriculture has become an increasingly capital intensive industry. The flow of capital into and out of agriculture depends on the cashflows generated and their associated risks, especially in relation to those of alternative investments. With the exception of a few dramatic but brief periods, attracting capital investment into U.S. agriculture has not been a significant problem.

Financial markets are also changing rapidly due to advances in technology, financial innovations, and deregulation. Many restrictions that contributed to financial market inefficiency in the past, such as interest rate ceilings, geographic limitations on bank activities, and limitations on permissible financial activities of commercial banks, have been or are being liberalized. Such changes increase financial market integration and competition, reducing local market power and dependence on locally generated investment funds. But such innovations and the recent rapid consolidation among financial intermediaries (Figure 6.4) give rise to concerns about the potential for increased market and political power and about a loss of local commitment and control, especially in smaller communities (see, for example, Gilbert 1997; U.S. General Accounting Office 1997). However, many believe small locally based lenders have competitive advantages that will allow many to survive and prosper despite increased competitive pressures prompted by financial market restructuring (Amel 1997). In addition, consolidation has not reduced the total number of banking outlets (Figure 6.4).

Past federal credit policies have been largely ineffective in supporting farm income, preserving family farms, overcoming discrimination, stimulating pro-

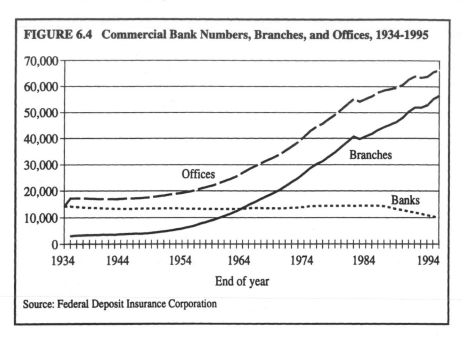

FIGURE 6.4 Commercial Bank Numbers, Branches, and Offices, 1934-1995

Source: Federal Deposit Insurance Corporation

duction or infrastructure development, or enhancing retail level credit market competition. Some success has been attained in reducing variation in the cost of credit across regions, providing liquidity to rural credit markets, and reducing credit rationing or redlining although the benefits of these successes have probably not always outweighed their costs.

Agriculture no longer dominates most rural economies, nor is it a sector characterized by widespread poverty. Substantial public benefits are probably not available from treating farm businesses or farm families differently from other small businesses or other families. A more compelling case can be made that removing federal preferences for agricultural credit without removing them from other favored sectors—housing, education, utilities, and small businesses— would cause economic distortions in the allocation of resources. That is, a second-best argument can be made about the level of distortions in each sector of the economy relative to other sectors (Hughes, Penson, and Bednarz 1984).

Given our review of past policies, their limited effectiveness, and changes occurring in agricultural production and financial markets, reasonable future federal actions with respect to agricultural credit markets may be substantially more limited than they have been historically. Here we suggest several general alternatives. They include lowering legal barriers to competition coupled with enforcement of antitrust laws, improving secondary markets, some changes in direct and guaranteed federal farm lending programs, and some possibilities for targeting credit without direct subsidies.

Lower Legal Barriers to Entry and Enforce Antitrust Laws. Perhaps the most compelling policy alternative for enhancing credit market performance involves continuing to remove artificial barriers to competition combined with enforcement of antitrust laws. Doing so helps create a business environment conducive to private competition, which in turn delivers benefits to market participants. Such policies have been undertaken in many heavily regulated sectors of the economy in recent decades, including transportation (rail, barge, truck, and airline), long distance telephone service, and natural gas production and transmission. Since the late 1970s, many such developments have been directed at financial markets, too. Changes in financial regulation include removing interest rate ceilings, allowing interstate banking and branching, allowing banks to offer insurance and mutual funds, and other initiatives.

Such policies encourage innovation, organizational flexibility, and efficiency, allowing for more rapid adjustment to new technology, changing demographics, and other market conditions. However, some deregulated markets have evolved in noncompetitive ways that partially thwart the intent of deregulation. A notorious example is the ability of one or two airlines to control immense shares of traffic at particular airports. Thus, deregulation may not automatically lead to sustained competitive business environments without ongoing antitrust vigilance.

Unfortunately, many rural financial markets are too small or diffuse to support a competitive array of locally based lenders. The potential exists in some rural areas and market niches for nontraditional and/or nonlocal competitors using communications technology to create competitive pressures once they gain acceptance from local consumers.

Improve Secondary Markets. Another policy alternative is to develop competitive secondary markets to provide liquidity to all lenders, including those serving agriculture. Currently, most GSE's provide liquidity to retail lenders by operating secondary markets for eligible loans or by taking such loans as collateral for advances. Secondary markets supply liquidity to local markets beyond that available from local deposits. In addition, secondary markets help integrate local credit markets with national money markets by facilitating the flow of funds among markets. By standardizing contractual arrangements for certain eligible loans and by providing a time series of consistent data by which to judge loan performance, GSE's have also helped attract additional capital to the sectors in which they are chartered to operate. Some secondary-market GSE's (Fannie Mae and Freddie Mac) have also enhanced retail-level competition for making eligible loans by encouraging new entrants, while others (Federal Home Loan Bank System) primarily benefit member institutions without diminishing their market power.

Competition could be further enhanced by harmonizing the charters of existing lenders to reduce market segmentation. Currently, the GSE status of FCS associations affords them advantages in providing larger, longer-term (mostly real estate) loans, while banks' ability to provide a full array of financial services as well as to accept insured deposits gives them a competitive advantage among small to midsized, largely part-time farms. In addition, exclusive charters and other factors prevent multiple FCS lenders from serving a given locality and reduce the competition-enhancing impact of FCS activity on local market competition.

Achieving maximum economic benefits from secondary-market activity requires vibrant competition at all levels of the intermediation chain. Otherwise, supranormal profits are captured (or operating inefficiencies persist) at those links where participants have market power. Policies that limit competition among GSE's allow them to make supranormal profits (or operate inefficiently), while policies that give preferential GSE access to particular lenders allow those lenders with market power to capture the benefits of that access rather than pass them through to retail borrowers. The major distinguishing factor among GSE's is the sector that they are chartered to serve; otherwise, most GSE's perform essentially the same function. Currently, each GSE is limited to a particular sector. This segmentation prevents GSE's from diversifying across industries, increasing their exposure to business risk and shielding them from competition with similar firms. This increased risk exposure also increases the contingent liability attached to the implicit federal backing associated with a GSE charter.

Change Direct or Guaranteed Government Lending. From an economic perspective, we see little use for traditional government credit programs. Herr (1994) noted there will always be excess demand for subsidized credit, but using subsidized credit to assist submarginal farmers is counterproductive. Such credit tends to slow adjustment, whereas policy should ease the exit of excess resources and labor to other industries, allowing farms to consolidate into viable units. To the extent that private markets may underfund beginning, minority, and limited-resource (low income, low equity) farms and to early adapters of new technology, federal policies that lower information and transactions costs may be appropriate. Guarantees substitute for information and can lower the lender's costs of making loans by shifting credit risk. Such guarantee programs can work if federal underwriting and monitoring standards and risk pooling are sufficient to control adverse selection and moral hazard problems. If market imperfections are present, sufficient guarantee fees can be collected to cover losses and administrative costs; and guarantee programs can be self-financing, as the Federal Housing Authority has demonstrated. Similar arguments can be made with respect to credit rationing. If credit rationing exists, borrowers should be willing and able to cover the cost of guarantee fees to overcome rationing and make the guarantee program self-financing.

Encourage Private Sector Lenders to Target Underserved Populations Without Direct Subsidies. Several studies (Beshouri and Glennon 1996; Canner and Passmore 1996) indicated that financial institutions that lend to lower-income clients (in the spirit of the Community Reinvestment Act—CRA) can earn returns comparable to those of other lenders. These results indicated that market inefficiencies may be present and that CRA-type regulation may help overcome them. The CRA requires regulators of depository institutions to encourage lenders to "help meet the credit needs of the local communities in which they are chartered consistent with ... safe and sound operations." Brewer and others (1996) found that Small Business Investment Companies (SBIC's) run by for-profit lenders tended to perform better than others. SBIC's are government-chartered private firms that provide long-term debt and equity financing to small firms financed by (implicitly subsidized) debentures guaranteed by the Small Business Administration. Thus, some market imperfections arising from discriminatory practices or information problems may be overcome by certain government actions in conjunction with private sector lenders.

Conclusion

The structure of federal intervention in agricultural credit markets is anachronistic. Agriculture is no longer a special case, and hence federal intervention via sector-specific credit programs and policies may no longer be warranted. Existing GSE's and other federal activity were established when 30 percent of the U.S. population was on farms, farms dominated most rural communities, and farm households were poorer than urban households. Now, farm households are a small percentage of total population and farms dominate fewer rural economies. Average farm household income is now equal to or higher than average urban household income and average farm household wealth exceeds the national average. In addition, the role of government support of agriculture is being reexamined. The Federal Agriculture Improvement and Reform Act of 1996 and recent changes in disaster relief and crop and revenue insurance increasingly shift responsibility for production decisions and risk management to producers, further reducing the logic of using federal farm credit programs as a safety net to replace lost incomes from economic emergencies or natural disasters.

As economists, we can assess the relative costs and benefits of policy alternatives and determine whether credit is the most appropriate intervention. Income redistribution programs are often disguised as market perfecting; and rent-seeking constituencies often find compelling social concerns to support their agendas. Credit programs justified on the basis of public benefits have a long history of being ineffective in achieving their goals. Direct transfer payments are more efficient to meet redistribution objectives. Where market failures or imperfections can be shown to exist, the cost-effective approach is to directly address their causes, for example, through antitrust policy or regulatory reform. When doing so is not feasible, unsubsidized loans or actuarially fair guarantee programs should be sufficient to overcome market failure or imperfections. Providing access, but not subsidies, also reduces crowding out and raises benefit/cost ratios.

References

Adams, D. W., D. H. Graham, and J. D. von Pischke, editors. 1983. *Limitations of Cheap Credit in Promoting Rural Development*, Economic Development Institute Training Materials CN-93, Washington, DC: International Bank for Reconstruction and Development.

Amel, D.F. 1997. "Antitrust Policy in Banking: Current Status and Future Prospects," Conference on Bank Structure and Competition. Federal Reserve Bank of Chicago. April 30-May 2.

Barry and Associates. 1995. *Agency Market Funds, Commercial Banks, and Rural Credit.* Champaign- Urbana, IL.

Berger, A., and T. Hannan. 1994. *The Efficiency Cost of Market Power in the Banking Industry: A Test of the 'Quiet Life' and Related Hypotheses*, Finance and Economics Discussion Paper 94-36. Washington, DC: Board of Governors of the Federal Reserve System.

Beshouri, C. P., and D. Glennon. 1996. "The CRA as 'Market Development' or 'Tax': A Test Using Credit Union Data," Conference on Bank Structure and Competition. Federal Reserve Bank of Chicago. May 1-3.

Bosworth, B. P., A. S. Carron, and E. H. Rhyne. 1987. *The Economics of Federal Credit Programs.* Washington, DC: The Brookings Institution.

Brake, J.R. 1974. "A Perspective on Federal Involvement in Agricultural Credit Programs," *South Dakota Law Review*, 19: 567-602.

Brewer, E., III, H. Genay, W. E. Jackson, and P. R. Worthington. 1996. "How Profitable is Small Business Financing? Evidence from Small Business Investment Companies in Financial Distress," Conference on Bank Structure and Competition. Federal Reserve Bank of Chicago. May 1-3.

Buttari, J. J. 1995. *Subsidized Credit Programs: The Theory, the Record, the Alternatives.* USAID Evaluation Special Study No. 75. Center for Development Information and Evaluation, U.S. Agency for International Development. June.

Calomiris, C. W., R. G. Hubbard, and J. H. Stock. 1986. "The Farm Debt Crisis and Public Policy," *Brookings Paper on Economic Activity*, 2: 441-479.

Canner, G. B., and W. Passmore. 1996. "The Financial Characteristics of Commercial Banks that Specialize in Lending in Low-Income Neighborhoods and to Low-Income Borrowers," Conference on Bank Structure and Competition. Federal Reserve Bank of Chicago. May 1-3

Carey, M. 1990. *Federal Land Banks, Market Efficiency and the Farm Credit Crisis.* Ph.D. Thesis, University of California, Berkeley.

Collender, R. N. 1991. "Production Economies and Inefficiencies Among Farm Credit System Associations." American Agricultural Economics Association Conference, Baltimore, MD, August.

Collender, R. N., and A. Erickson. 1996. *Farm Credit System Safety and Soundness*, AIB-722. Economic Research Service, U.S. Department of Agriculture, Washington, DC.

Dodson, C. B. 1997. "Production Contracts and Debt Usage on Commercial Farms" in Ellinger, P. N. (ed.), *Regulatory, Efficiency, and Management Issues Affecting Rural Finance Markets*, Staff Paper ACE 97-02, Urbana, IL: Dept. of Agriculture and Consumer Economics., University of Illinois, June, pp. 110-124.

Dodson, C. B. 1996a. "Analysis of Lender-Borrower Choice and Implications for Federal Farm Credit Policy," in *Regulatory, Efficiency, and Management Issues Affecting Rural Financial Markets*, Staff Paper SP0196. Fayetteville, AR: University of Arkansas, pp. 149-174.

Dodson, C. B. 1996b. "The Changing Structure of Nonreal Estate Credit" in *Agricultural Income and Finance*, AIS-60. Economic Research Service, U.S. Department of Agriculture, Washington, DC, pp. 33-40.

Dodson, C. B. 1996c. "A Multinomial Logit Analysis of Mortgage Borrowers By Lender Group," American Association of Agricultural Economics Association Conference, San Antonio, TX, July.

Dodson, C. B. 1996d. *Is More Credit the Best Way to Assist Beginning Low-Equity Farmers?* AIB 724-04. Economic Research Service, U.S. Department of Agriculture, Washington, DC.

Dodson, C., and S. Koenig. 1994 "The Major Farm Lenders: A Look at Their Clientele," *Agricultural Outlook*, AO-214, Economic Research Service, U.S. Department of Agriculture, Washington, DC. December, pp. 24-27.

Dodson, C., and S. R. Koenig. 1995a. "Young Commercial Farmers: Their Financial Structure and Credit Sources," in *Agricultural Income and Finance*, AIS-56. Economic Research Service, U.S. Department of Agriculture, Washington, DC, pp. 40-44.

Dodson, C. B., and S. R. Koenig. 1995b. "Niche Lending in Agriculture," *Journal of Agricultural Lending*, 8 (4): 14-20.

Duncan, M. 1997. "Keynote Address: Where Are Rural Capital Markets Headed?" *Financing Rural America*, Kansas City, MO.: Federal Reserve Bank of Kansas City, pp. 11-46.

Farm Credit Administration. 1995. *1995 Annual Report on the Financial Condition and Performance of the Farm Credit System*. McLean, VA.

Farm Credit Banks. 1995. *Young, Beginning, and Small Farmers and Ranchers Reports*. Unpublished reports submitted to the Farm Credit Administration.

Gale, W. G. 1991. "Economic Effects of Federal Credit Programs," *American Economic Review*, 81: 133-152.

Gardner, B., 1992. "Changing Economic Perspectives on the Farm Problem," *Journal of Economic Literature*, 30: 62-101.

Gilbert, R. A. 1997. "Implications of Banking Consolidation for the Financing of Rural America," *Financing Rural America*, Kansas City, MO: Federal Reserve Bank of Kansas City, pp. 131-140.

Glauber, J. W., and M. J. Miranda. 1996. "Price Stabilization, Revenue Stabilization, and the Natural Hedge," unpublished manuscript, 30 pp., October 30.

Herr, W. M. 1994. "Are Farmers Home Administration's Farm Loan Programs Redundant?" *Agricultural Finance Review*, 54: 1-14.

Hoppe, R. A., R. Green, D. Banker, J. Kalbacker, and S. Bentley. 1996. *Structural and Financial Characteristics of U.S. Farms, 1993, 18th Annual Family Farm Report to Congress*, AIB 728, Economic Research Service, U.S. Department of Agriculture, Washington, DC, October.

Hughes, D. W., J. B. Penson, Jr., and C. R. Bednarz. 1984. "Subsidized Credit and Investment in Agriculture: The Special Case of Farm Real Estate," *American Journal of Agricultural Economics*, 66: 755-760.

Johnson, D. G. 1963. "The Credit Programs Supervised by the Farm Credit Administration," in *Federal Credit Agencies*, U.S. Commission on Money and Credit. Englewood Cliffs, NJ: Prentice-Hall, Inc., pp. 259-318.

Kalbacker, J., and D. Rhoades. 1993. "Profiling Black Farmers in the United States," *Agricultural Outlook*, Economic Research Service, U.S. Department of Agriculture, Washington, DC, December, pp. 25-29.

Keeton, W. R. 1995. "Multi-office Bank Lending to Small Businesses: Some New Evidence," Federal Reserve Bank of Kansas City, *Economic Review*, 80 (2): 45-57.

King, R., and R. Levine. 1993. *Finance, Entrepreneurship, and Growth: Theory and Evidence*. World Bank Conference: How Do National Policies Affect Long-run Growth? January.

Koenig, S. R., and C. B. Dodson. 1995. "Comparing Bank and FCS Farm Customers," *Journal of Agricultural Lending*, 2 (2): 24-29.

Koenig, S. R., and J. T. Ryan. 1997. "Farm Loan Volume Prospects for the New Farmer Mac." in Ellinger, P. N. (ed.), *Regulatory, Efficiency, and Management Issues Affecting Rural Finance Markets*, Staff Paper ACE 97-02, Urbana, IL: Dept. of Agriculture and Consumer Economics., Univ. of Illinois, June, pp.58-78.

LeBlanc, M., and J. Hrubovcak. 1986. "The Effects of Tax Policy on Aggregate Agricultural Investment," *American Journal of Agricultural Economics*, 68: 767-777.

Rhyne, E. H. 1988. *Small Business, Banks, and SBA Loan Guarantees: Subsidizing the Weak or Bridging a Credit Gap?* New York: Quorum Books.

RUPRI Rural Finance Task Force. 1996. *Seven Policy Issues on Financial Markets and Rural Economic Development*. Columbia, MO: Univ. of Missouri, Rural Policy Research Institute.

Stam, J. M. 1995. *Credit As a Factor Influencing Farmland Values*, ERS Staff Paper 9504. Economic Research Service, U.S. Department of Agriculture, Washington, DC.

Stam, J. M., S. R. Koenig,, S. E. Bentley, and H. F. Gale, Jr. 1991. *Farm Financial Stress, Farm Exits, and Public Sector Assistance to the Farm Sector in the 1980's*, AER-645. Economic Research Service, U.S. Department of Agriculture, Washington, DC.

Sullivan, P. J. 1993. "The Structure of Bank Markets and the Cost of Borrowing: Evidence From FmHA Guaranteed Loans," in *Regulatory, Efficiency and Management Issues Affecting Rural Financial Markets*, Staff Papers Series 93-22. Tallahassee, FL: Food and Resource Economics Department, Univ. of Florida, pp. 178-205.

U.S. Council of Economic Advisors. Various years. *Economic Resport of the President*. The White House, Washington, DC.

U.S. Department of Agriculture. 1997a. *Agricultural Income and Finance*, AIS-64. Economic Research Service, Washington, DC, February.

_____. 1997b. *Civil Rights at the United States Department of Agriculture: A Report of the Civil Rights Action Team*, Washington, DC.

_____. 1997c. *Credit in Rural America*, AER 749, Economic Research Service, Washington, DC.

U.S. General Accounting Office. 1997. *Rural Development: Availability of Capital for Agriculture, Business and Infrastructure*. Washington, DC

U.S. Department of the Treasury. 1996. *Government Sponsorship of the Federal National Mortgage Association and the Federal Home Loan Mortgage Corporation*. Washington, DC, July 11.

7

Emerging Nontraditional Lenders and Products

Bruce J. Sherrick

Nontraditional lenders are defined as either those whose historic contacts with farm borrowers were for products or services other than debt, or whose current product offering is markedly different from debt capital products traditionally supplied to agriculture. Examples include vendor finance operations, leasing companies, specialized resource "bundlers," equity packagers (venture capital and investment banking), providers of private enhancements to support debt, guarantors from agribusiness, as well as traditional lenders offering specialized products or services to their customers.[1] Industry observers have noted the increased prominence of these lenders and their products, and have raised questions about their motivations, performance, and impacts on both their borrowers as well as traditional lenders. This chapter begins to address some of those questions by examining emerging nontraditional lenders and their products, with particular attention to the non-real estate capital markets. At the same time, other questions are raised about industry and market forces on which the future of these lenders depends.

The specific purposes of this chapter are to:

- provide economic rationale and an organized "framework" to understanding the motivations and performance of nontraditional lenders;
- examine and evaluate specific nontraditional lenders and their products;
- help formulate expectations about the future role of nontraditional lenders in changing markets for agricultural capital.

The chapter begins by providing brief evidence about the size and coverage of nontraditional lenders. Anecdotes about their performance—and commonly cited reasons for their existence—motivate a discussion of the intermediation functions they perform. The chapter then provides a nontechnical interpretation of recent advances in the literature that can help provide a framework to assess the economics of intermediation by nontraditional capital suppliers. It then examines competitive implications of the presence of nontraditional lenders for the industry and customers. Throughout, product innovations and the role of "unbundled" financial products in meeting the needs of ever more specialized borrowers are discussed. The chapter concludes with some views on the future—not to predict its form *per se*, but to highlight rapidly changing market segments that deserve future attention, and to highlight critical issues that will determine the future role of nontraditional lenders.

Evidence of Size and Coverage

Unlike most commercial lenders who are required to report conditions and activities to their regulators through regular Call Reports, there is no organized reporting mechanism for most nontraditional lenders. Thus, their activities must be examined in less direct fashion—either by observing borrowers and their reported debt positions, or by surveying nontraditional lenders and making inferences about the behavior of the group to which they belong. USDA collects extensive information about sources and levels of operator debt through the Farm Cost and Returns Survey, from which general indications of market share can be gleaned. From this source, Dodson (1996) reports, "For commercial sized farms, nontraditional lenders represent the second largest source of debt." Although specific subcategories are not reported, individuals and others—including trade and vendor credit—steadily increased market shares of non-real estate debt since 1987 and represented 24 percent of the total, or $17.5 billion, in 1996 (Figure 7.1). Of the subset of farms using trade credit, manufacturers and dealers accounted for 59 percent of their total non-real estate debt (Dodson 1996). From the other side of the market, surveys of equipment manufacturers with captive finance programs and major leasing companies indicate that the majority of this debt is for farm machinery and supplies financed through operations such as Deere Credit or J.I. Case Credit (Sherrick et al. 1997).

The increasing market share captured by nontraditional lenders is further indicated in case studies of specific firms such as AgServices of America, a company started in 1987 to help farmers purchase and finance operating inputs. In its first decade, AgServices posted an annualized growth rate of approximately 50 percent. During the last 5 years, it expanded coverage to 26 states, nearly tripled its number of customers, and moved to wholesale, asset-backed securitization for funding. And the company feels that they are currently at 1 percent of their long-run market size. Likewise, their customers indicate high levels of satisfaction, and their product offerings continue to expand rapidly into related financial products and management services (URL: *www.agservices.com*). This favorable per-

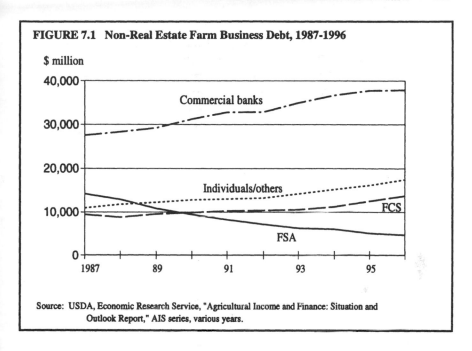

FIGURE 7.1 Non-Real Estate Farm Business Debt, 1987-1996

Source: USDA, Economic Research Service, "Agricultural Income and Finance: Situation and Outlook Report," AIS series, various years.

ception by customers is steadily replacing a view, historically more common, that use of "storefront" finance signaled less creditworthiness and an inability to secure traditional financing.

The most visible nontraditional lenders are probably the equipment manufacturers' captive vendor finance operations. Although companies such as Deere Credit have existed in some form for many years, their activities are nonetheless classified as those of a nontraditional lender as well. And, as with GE Capital Corporation when it moved to financing products beyond its own production channels, these nontraditional lenders in agriculture have begun "second generation" activities that compete in traditional financial domains. Deere, specifically, has recently moved beyond direct vendor financing of products at their own dealerships to financing vendor operations at wholly unrelated companies.[2] Even traditional lenders are beginning to venture into partnerships with manufacturers to provide vendor finance facilities. The Farm Credit System (FCS) has instituted a national Trade Credit Desk, and many of the associations have developed their own versions to fund the operations of nontraditional lenders, or provide two-desk operations to originate and underwrite credit at the point of sale. These hybrid operations resemble early developments of acceptance corporations that have matured in automobile lending. It remains to be seen if the vendor finance operations in agriculture will develop as extensively as in automobile lending, and whether other forms of intermediation will evolve. In any case, the evidence, even if largely anecdotal, suggests that these nontraditional lenders are gaining prominence and visibility as they add to their collective volume.

Commonly Cited Reasons for Existence
of Nontraditional Lenders

Explanations for the existence of nontraditional finance operations include: (1) product marketing motivations; (2) filling otherwise unmet demands (shortfalls in credit supply); (3) promotion or protection of proprietary technologies or specialized assets; (4) development of supply arrangements (and prevention of supply uncertainty); (5) facilitating new entrants in agriculture; (6) exploitation of a profitable operating center; (7) reaction to, or distinction from competition; and (8) expansion of a parent company's effective borrowing capacity.

Product Marketing Motivations

Each of the manufacturers with vendor finance operations surveyed in Sherrick et al. (1997) cited the marketing benefits of offering credit as a primary motivation for operating their financing facilities. Whether the finance operations were profitable on its own or not, manufacturers viewed the provision of credit as a mechanism to increase sales on which they then earn manufacturing or sales margins—generally greatly in excess of the margin earned on the lending operation itself. Of course, for this effect to hold, the provision of vendor credit must result in additional sales rather than the simple substitution of financing on sales that would have taken place anyhow. Representatives at Growmark, a major Midwest supply cooperative, estimated that the financing provided had a net additionality of sales of 20-25 percent, and that average margins on products sold were nearly 20 percent. And, even without additional sales, for many manufacturers, it is better to store receivables on the balance sheet than inventory.

Low interest rates, or otherwise subsidized credit, have frequently been used by equipment manufacturers to stimulate sales. The motivations may include attempts to smooth inventory levels and stabilize production cycles, or simply to manage cycles of capital replacement that are derived from farm income cycles. In any case, investigations of the no-interest-for-6-months credit program at Deere revealed that the marketing department for Deere pays the financing unit for the foregone interest charges. The marketing department apparently views credit as an effective form of promotion. This particular case also illustrates the complexity of evaluating the credit function on its own when its use is intertwined with marketing efforts.

Unmet Demands, or Shortfalls in Supply

Manufacturers also provide credit or credit enhancements in order to fill perceived credit gaps, and to help smooth out the cycles in supply of credit, particularly in small, localized markets. While there has been little empirical evidence of credit shortfalls in agriculture, some borrowers probably feel that vendor finance operations make credit more available than traditional lenders. The appropriate underwriting standards of nontraditional lenders with manufacturing margins at stake and different collateral risks can be quite different than those of traditional lenders with only an interest margin at stake. Thus, the perception that nontraditional lenders are filling credit gaps may be a manifestation of differing credit standards rather than actual shortages in available risk-priced credit.

On the equity side, however, standardized markets have not developed in agriculture that permit leverage decisions to be routinely considered independently of size or scale. Thus, in terms of providing venture capital, subordinated equity or debt, and investment banking services, nontraditional providers may in fact be filling "gaps" in the market.

Promotion or Protection of Proprietary Technologies or Specialized Assets

Traditional lenders are sometimes unprepared to evaluate highly proprietary technologies or specialized assets, or are unwilling to assume the contingent liability of failure in cases where there is not good information about the likely performance of the firm. In these cases, vertical coordination through financing by an integrator or output processor, and contingent equity ownership as conveyed through security agreements, can be an appropriate means of capitalizing production units. For example, milk tank manufacturers were the first to finance the acquisitions of onfarm milk cooling and storage systems until their efficacy was proven. Similarly, integrators provided much of the capital needed to expand hog facilities to a scale that permitted the exploitation of new health management and feeding technologies. There can be good reasons for nontraditional financing of changing production technologies and of unique risks. Thus, to the extent that the innovations involving specialized assets or technologies continue or accelerate, the need for specialized provision of capital to support these assets will likewise continue or accelerate. Increased specialization of financial products—frequently identified as nontraditional lending—is the natural result of evolution in the production sector.

Development of Supply Arrangements, and Management of Supply Uncertainty

In contrast to, say, the existence of specialized assets that preclude a willing traditional lender, there may be easily "bankable" situations for which a nontraditional lender would still naturally emerge instead to solidify supply arrangements. For example, many marketing contracts serve to alleviate some risks of the processor as well as ensure minimum prices for the producer. The benefit sharing that takes place under these contracts can be transmitted in the form of inexpensive financing provided by the output processor, or as an enhancement by an input supplier, or in many other forms that are easily denominated in the price of capital. Thus, financial terms may simply be altered to convey the benefits directly from one party to the other when financing is involved, thereby avoiding a third-party financier through whom both accounts would otherwise have to be cleared.

Facilitating New Entrants in Agriculture

Nontraditional financing often fills new entrants' startup needs, especially when traditional underwriting results in credit rationing. The most common form occurs with subordinated debt, or quasi-equity provided by an integrator. Government credit programs that provide preferential treatment to startup or young operators could likewise be classified under this heading. The argument, in any case, is that traditional lenders concerned only with the interest margin find it impossible to risk-price and still provide positive funds. The motivation for governments to provide this type of program is either to serve some ancillary objective or to foster a particular organization of production units that is publicly preferred.

Exploitation of a Profitable Operating Center and Wholesale Credit Efficiencies

In this case, the distinction from a traditional lender is somewhat lost if the functions of a traditional lender can be performed profitably *on their own* by a nontraditional. Several large vendor finance operations in agriculture have at times accounted for large portions of the parent companies' net incomes.[3] In these cases, the challenge is to explain how or why there are any advantages for the "nontraditional" to perform traditional functions. For now, the point is that nontraditional lenders have emerged with competitive advantages in traditional lending markets.

In cases where there are numerous developed vertical relationships, an integrator as a provider of financial capital conceptually "collects" production units into a single operation and, in effect, borrows on behalf of the collective producers. Thus, if there are advantages in "wholesale" borrowing on behalf of a set of coordinated production units, it would be expected that the nontraditional arrangement would result. In other cases, where the contractual linkages are less clear, the nontraditional supplier may still change the nature of the credit used in terms of amount and type. In other words, the fungibility of funds may allow a producer to finance feed, for example, on favorable terms with a feed miller, thereby freeing cash and avoiding the need to finance purchases of the products provided by the integrator; or it may result in changing the term of the debt by eliminating the need to finance short-term assets and increasing the commitments of longer term assets. The total debt in the production system is unchanged, but rearranged to parties with the lowest economic costs or greatest efficiencies in risk-bearing.

A Reaction to Competition

Researchers at the University of Illinois recently conducted a privately commissioned study of the provision of credit by a seed corn company (Sonka et al. 1997). The study was done at the behest of Context Consulting, who were considering the addition of a financing unit at a seed corn company undertaken almost entirely because a competitor had done so. The contracting seed company was interested in removing any distinction or advantage that the existence of a financing unit by the other seed company might convey. While the specific details are not public, the study concluded near-indifference by customers, or even some suspicion that use of such trade credit might signal less creditworthiness. Nonetheless, many input suppliers continue to look for ways of providing trade credit in conjunction with their products for the purposes of "keeping up" with competitors.

Expansion of a Parent Company's Effective Borrowing Capacity

The finance literature contains mixed empirical evidence as to whether the formation of a captive finance company effectively increases the borrowing capacity of the parent company (Fooladi et al. 1986; Roberts and Viscione 1981a/b). The financing company, by its nature, can operate at higher levels of leverage than can the production unit, and all it really does is carry the same receivables (or inventory conversions or inventory factoring to parent), thereby resulting in effective total leverage increases. Under this justification, the lending operation itself is somewhat incidental to the purposes of the parent company, but displaces traditional lenders nonetheless.

Intermediation Functions of Conventional Versus Nontraditional Lenders

The previous section was meant to familiarize the reader with popularly cited reasons for nontraditional lenders, not to argue for or against legitimacy in specific cases. In what follows, the lending function is distilled into its component parts to further highlight the differences between conventional and nontraditional lenders.

Funding

Funding costs account for the single largest component in the cost of extending credit for most lenders. Traditional lenders utilize traditional funding sources —banks have depository authority, access to wholesale lending markets, and in some cases, indirect advances from agency market sources. The FCS, as another traditional lender, has a funding corporation that sells systemwide government agency-status bonds in wholesale debt markets, and can accumulate member capital to help fund its assets. While there is some evidence and concern over "deposit drain" and loss of local core funding sources to brokered money market funds and the like, traditional lenders have for the most part enjoyed fairly inexpensive and stable funding sources. Further, they have been participants in many of the notable financial innovations and financially engineered products that have been developed to manage interest rate risk and their investment portfolios. Nontraditional lenders, by contrast, have no authority to accept deposits, nor do they share in the potential advantage of having a unified funding agency with quasi-agency status. Instead, they must use other sources of funds or develop new mechanisms for funding lending operations—and they have done both.

Among existing sources are commercial paper markets (especially suitable for large operations like Deere), sources on the parent company's balance sheet, correspondent banking relationships, and use of traditional lenders to fund the lending operations of a nontraditional lender. For this latter type of re-lending to make sense economically, the benefits of bulk borrowing and other advantages the nontraditional may have (in terms of access to customer information and so on) must outweigh the disadvantages of an additional transaction and the margin the original lender would need to earn on its loan to the nontraditional.

An additional avenue for funding, securitization, is the creation of standardized securities representing equivalent claims against a set of uncertain cash flows. Common applications include firms such as Farmer Mac that buy loans, package them, and sell standardized claims against the set in total. The benefits of diversification can be an important motivation, as well as matching cash-flow profiles (and new denominations of value) to investors whose preferences may not match the characteristics of the original debt obligations. Among manufacturers or vendors, securitization provides a funding mechanism for a lending unit to securitize and sell its accounts receivable. Although the recent growth in this practice has slowed, it remains a viable option, particularly for nontraditional lenders with long histories and large volumes. Companies such as Capital Markets Acceptance Corporation (CapMac) and First Chicago are driving down the minimum size required for securitization to be an efficient practice (now reportedly below $20 million at First Chicago), and it is likely that the minimum-sized transaction will continue to fall over time. Thus, securitization is likely to become a more accessible source of funding for some nontraditional lenders through time.

Simply being a nontraditional lender does not automatically convey funding advantages or disadvantages, but the organization and operation of some of the nontraditional lenders does result in funding sources that are closely matched to the lending products offered. Further, the total costs of lending depend only partly on funding costs, so the advantages of one type of lender over another in this category of costs may be more than offset by other cost differences.

In addition to direct debt products, capital providers to agriculture are competing against other opportunities for the employment of venture capital. Direct equity markets in agriculture have been somewhat slow to develop, but the continued integration in livestock has begun to generate cases where coordinated capital syndicates have been assembled involving numerous types of lenders and financial products. Cases where the funding is specifically matched to the type of operation being funded frequently require participation of nontraditional lenders or at least nontraditional products by traditional lenders. Finally, to the extent that interest rate risk is a modifier of funding costs, size and experience of the lender are likely positively related to the agility with which this risk is managed.

Origination and Delivery

Henricksen and Boehlje (1995) report that captives (as one form of a nontraditional lender) have significant cost advantages over traditional lenders in making smaller loans (under $50,000), and marginal advantages in larger loans as well as in working out problem loans. Part of the advantage arises from the ability to share "brick and mortar" expense with existing business operations, and part from the fact that they spend less time than traditional lenders. They may spend less time directly on the loan application process because they already know more about the customer by virtue of the other business conducted with that customer, or because they simply collect and analyze less information. If one includes costs born by the borrower as well, then point of sales financing earns the additional advantage of one less location/transaction for the customer in arranging financing that otherwise would be separate from the purchase decision. In any case, the rigidity and traditions of delivery of credit through traditional channels is likely a disadvantage for traditional lenders relative to more flexible nontraditional lenders.

Underwriting and Credit Risk Assessment

Experience and capacity for formal evaluation of credit are likely to be greater at a traditional lender. Nontraditional lenders have favored less extensive and more generic information (credit bureau reports, scoring approaches) and rely on security agreements and assignments to minimize risk. Rather than develop their own capacity to perform the underwriting functions, some nontraditional lenders (and some traditional lenders as well) have moved to third-party vendors of these services. As information systems become more sophisticated, the underwriting function becomes more separable. And, as this particular function becomes more separable from the other functions of lending, differences in access or quality by lender type become less obvious. Thus, the location at which this particular function is housed (within a traditional lender's operation or at a standalone third party or at the nontraditional) becomes innocuous to the performance of the capital market it serves.

Credit Risk Bearing

Closely related to the competence of credit risk assessment is the *capacity* to bear the risks of loss associated with nonperformance by the borrower. The contrasts between traditional and nontraditional lenders are probably more acute in this case. First, the collateral valuation may be different by lender. For example, a repossessed tractor may be of greater value to a vendor finance unit of an equipment manufacturer than to a local bank. There is even some folklore that a bank "looks" worse and suffers greater reputational costs when it repossesses collateral than does a vendor for whom the reacquisition of the equipment is less obvious. And if there are product margins at risk as well as lending margins, the vendor will rationally bear more credit risk than a traditional lender and will lend to riskier customers than will bank-only financing. A perennial complaint of traditional lenders is that nontraditionals can "cherry-pick" and move into and out of markets due to the flexibility afforded by having multiple lines of business. By contrast, lenders whose only line of business is lending are less able to move into and out of markets as conditions and signals warrant. Conversely, a financing subsidiary is vulnerable to the performance of the remainder of the company and may be cut or expanded for reasons unrelated to its own performance. The metaphorical financing tail may also wag the parent dog at times—as has been seen during the maturation of vendor finance operations in automobiles and in the large capital corporations that grew from within manufacturing parents.

Bonding Functions

Bonding functions relate to the provisions, both explicit and implicit, that affect the behavior of the borrower as a condition of financing. Traditional lenders are restricted in the degree of involvement they can engage in without becoming liable for adverse outcomes. Nontraditional financial contracting, on the other had, is often warranted explicitly *because* of the need for more direct involvement by the provider of financial capital or because of equity-like sharing of the benefits and/or costs. These differences become more pronounced as the financial products tend toward equity investments. Finally, loyalty between the customers and the lenders can influence performance of the loan in times of financial distress, although it is not clear how loyalty differences depend on the type of lender.

Warehousing

As it becomes easier to take loan assets off balance sheets by traditional lenders, the warehousing function becomes less distinct by type of lender. Like underwriting functions, it appears that the options have increased for managing the size of the holdings of loans assets, and thus both kinds of lenders enjoy similar opportunities for similar types of loans. However, nontraditional lenders will tend toward less homogenous products and more unique risks and thus may have to pursue more complicated means of selling their loans or receivable streams. Traditional lenders perhaps face more rigid overall size constraints given formalized capital regulations, but overall, the warehousing function does not appear to be a source of significant advantage by type of lender.

Servicing and Monitoring

Servicing of loans is becoming a more separable function of lending as well, so type of lender does not carry with it an automatic advantage. Traditional lenders have greater existing infrastructure, and one might expect this function to be housed proximate to traditional lenders' operations. In either case, servicing is becoming more of a fee-for-service operation with similar costs for whomever uses the service. The nontraditional lenders, particularly at point of sale, will tend to have better customer contact opportunities and thus will find the monitoring easier.

Collection and Workout

As reported by Henricksen and Boehlje (1995), cost rates for workout of problem loans are apparently lower at nontraditional lenders, but they will likely have more of it to do! If there are product margins involved or special risks making the loan less traditionally bankable, then the rate of failure by nontraditional capital providers would be expected to be higher. But, the closer relationships with borrowers and the relative ease of collateral disposal can offset the greater frequency with which nontraditionals encounter problem loans.

Regulatory Burdens

Banks are regulated at numerous levels by the Federal Deposit Insurance Corporation (FDIC), the Comptroller of the Currency, the Federal Reserve System, state and federal banking regulations, indirectly by Congress, and through the influence of the organized reporting procedures (Call Reports). Depending upon the organization, they may also be regulated by the Securities Exchange Commission, and indirectly by security rating agencies. The FCS is overseen by the Farm Credit Administration and indirectly by Congress. These relationships ostensibly exist to protect the safety and the soundness of the U.S. money and banking systems.

Although nontraditional lenders are also bound by property, lien, usury, and security laws, they operate in relative autonomy from formal regulators who can impose capital standards, loan quality ratings, and the like. This can be viewed as a potential advantage that offsets the firm's exclusion from deposit markets and other sources of wholesale funds. Although it is difficult to quantify the direct regulatory expenses for traditional lenders, they may be significant for some types of lenders (FDIC Call Reports). Further, to the extent that the equity of the nontraditional supplier is publicly traded, the "market" is a regulator if it in fact provides the necessary discipline for sound management of the lending operation.

Economic Approaches to Understanding
Nontraditional Lender Behavior

The previous sections discussed the various lending functions to help delineate differences across lenders. However, the lending functions themselves do little to provide a set of unified economic principles behind the observed market structure and its apparent continued evolution. Below are three general themes from the economics literature that may provide additional insights.

The first relies on the characterization of the *profit function* or similar objective function of the participants. In the case of vendor finance operations, Sherrick and Lubben (1994) present a simple model that can be used to compare different lenders in terms of their optimal risk exposure, number of loans, relative loan rates, and impacts of effective collateralization and other cost measures. The results indicate that lenders will find it attractive to lend to the point that the marginal expected costs of default equal the marginal benefit of lending—not a surprising result. But formalizing the model to include product margins and differences in collateral rates highlights the advantage of manufacturers in lending to buyers of their products, a result that is not apparent if one considers only the individual functions of lending or the differences between types of lenders at the same level of behavioral variables. The equilibrium behavior of different lenders depends on their entire portfolio and not just on the direct lending function alone. Because of the margin at stake in the product manufacturing, the vendor can profitably lend at lower interest rates than a bank, and can use the terms of financing to more effectively discriminate among different kinds of borrowers, thereby reducing the problems of asymmetric information about borrower conditions. The "scope" economies conveyed by the addition of lending to their other lines of business results in different "optimal" behavior than that of a traditional, one-line-of-business lender. Thus, relative to a bank, the vendor financier will employ underwriting standards (to segregate acceptable from unacceptable loans) that systematically differ from those the bank would find optimal. Nontraditional lenders will be able to make loans that the traditional lender would find unacceptable, either because of uncertainty about repayment or because of different consequences under the terms of default. This approach and other classical profit function models of firm behavior can assist in understanding behavior of specific firms in specific cases, but are difficult to aggregate into models of industrial organization.

Next, the *principal-agency* literature has received renewed attention with the insights it provides into the use of different contractual linkages to more nearly align the incentives of the participants. Several authors have developed explanations that posit the formation of captives as a rationale for reducing agency costs (Dipchand et al. 1982), and others have developed explanations of indirect lending as a mechanisms for reducing overall market screening costs (Staten et al. 1990). Both branches in this literature stress different information costs by different lenders and the possibilities of asymmetric information about borrower conditions. In general, this now-extensive literature holds promise as a means to identify contractual terms for sharing risks under financial contracts that avoid agency costs and improve total welfare of the participants. Agency costs arise from the different objectives of the lender and borrower, and nontraditional financing could be seen as more readily customized to accommodate the differences than some traditional forms of lending. In integrated or contracted production, for example, the traditional underwriting standards may not be appropriate to judge credit risk, and nontraditional lenders or nontraditional enhancements may be required to make the package acceptable to finance. This outcome can be seen as an implication of the principal-agency literature's guidance toward contracting, which results in placing the risks where most efficiently borne.

The third conceptual framework that assists in understanding emerging lenders and their products is known as the *state-space* framework. Despite its esoteric sounding label, it provides a rich framework to examine product innovations and determine their contribution to a market's ability to span risk dimensions previously uncovered. These concepts are employed extensively by insurers and were brought to prominence by financial engineers who used these approaches during the era in which components of interest rate contracts were repackaged and sold under myriad configurations. The state-space approach helps determine when different configurations of contracts are in fact redundant and add nothing to the market's ability to span and bear risk, and when they are in fact important innovations that increase the value of the parts from which they were constructed. In the case of interest-rate market innovations, the first interest-only (IO) and principal-only (PO) strips of treasuries were believed by many to simply be inefficient repackaging of existing claims. By analogy, early innovations by nontraditional lenders or repackaged financing could have been viewed as just another layer of costs added to an existing market. As the market demonstrated an enormous appetite for customized interest rate contracts, it became apparent that additional separation of the components was in fact serving a purpose—or "spanning" a dimension of the market, in the vernacular of the state-space models. There now exist a dizzying number of customized financial products based on claims against separated streams of Treasury bond payments alone. Other innovations—including swaps, customized options, and over-the-counter exchanges of contingent positions—are simply the result of rearranging the parts of existing marketable securities. Again, by analogy, developments in agricultural finance markets could likewise be viewed as simply rearranging the roles and parts of existing products and players. The state-space approach helps to answer *which* of the possible combinations or positions will add value to the market or provide a lower cost means to achieve similar risk-reward positions.

The essence of the state-space theory is that there are different possible outcomes in the future that can be collected or partitioned into sets of claims against which current prices are set. In the case of a lending contract, there are numerous levels of future firm performance that determine the performance on the financial obligations of the firm. In typical debt contracts, any revenue in excess of that needed to service the debt accrues to the equity holder and, under default, the lender gets whatever the value of the security is —essentially partitioning the outcome possibility set into two regions of default or nondefault. Firm performance, however, is a many-dimensioned function that could be described along axes including things like yield, costs, prices of outputs, weather, management, and so on. To be risk-efficient, the market must be able to ensure against any particular combination of the input variables that determine firm performance, but dividing the outcome space into only two partitions cannot "span" or replicate or efficiently price the risks involved with lending to that customer.[3] Simply dividing the outcome space into more independent regions permits more efficient pricing of the risks of lending to that customer. What may appear as simply creating redundant claims by reinsurers against certain outcomes can actually add to the total dimensions of risk spanned by the market instruments. In the case of agricultural lending, more numerous financial products that simply divide up the claims against the same set of outcomes in linearly independent combinations add to the ability to efficiently price risk. Even if there is no change in the aggregate supply of capital, the reallocation among the parties in the system results in greater "value" to the system. Under this paradigm, markets evolve to the point that either all relevant risk dimensions are spanned (i.e., any risk-reward profile can be obtained at some cost through reorganizing existing assets) or to the point that the additional benefit of more finely divided contracts is less than the additional cost of accounting for the different claims separately. In the end, risks are borne by those who can do so at least cost—and there is little reason to suspect *a priori* that traditional lenders operate at least cost . Thus, innovations in financially engineered claims in agriculture would likewise not be expected to develop exclusively in the venues of the traditional lenders. Again, by way of reference to interest-rate markets, agricultural markets may still be in the infancy of separating functions and claims into more and more specialized parts.

Sample of Nontraditional Lenders
and Products of Interest

The most visible and prominent of the "new" lenders are now profiled to illustrate some of the concepts and principles described earlier and to suggest future roles they may serve in financing agriculture. The list is by no means exhaustive, and the list of financial products they provide may soon be out of date.

Captives and Vendor Finance Companies

Both by extrapolating recent trends and by drawing analogies from other industries' experiences, the expectation is for continued growth in financing originated at point of sale or in conjunction with product provision. Deere has become one of the 25 largest financial institutions in the country, and companies like GE Capital Corporation and the big three automakers' acceptance and leasing corporations continue to grow at rates far exceeding the banking institutions. But what relationship will captives forge with traditional suppliers and how will the functions of lending be divided among the players? In any case, the vendors and captives with marketing motivations for providing credit appear to be here to stay. The advantages conferred by proximity to the transactions requiring financing, potential information advantages, and those of bonding and risk bearing, can perhaps be exploited while relying on fee-for-service provision of the other functions of lending. Recent examples of traditional lenders forming trade credit operations suggest that at least some lenders have figured out how to exploit their comparative advantages while offloading the functions that they may be at a competitive disadvantage in providing.

There is some concern about the stability of credit provided through nontraditional channels. By contrast, part of the traditional lending structure is based on a requirement to serve all segments of the market in good times and poor. Flexibility to serve or ignore a market is a valuable attribute to nontraditional lenders, but its existence can also destabilize the market to some degree.

The aggregate supply of capital is also probably enhanced by the existence of these lenders, but there has been only sparse evidence that there have been real shortages. The multitude of forms that captives and vendors use to convey capital, however, may improve the ability of many producers to manage their capital structure more easily.

Finally, the absence of regulatory impediments, and the ability of captives to change their behavior to quickly match new needs of their customers, will likely raise additional policy concerns. Traditional lenders may strive for additional authorities to compete, and nontraditionals may seek better access to funding channels reserved for government-chartered lenders.

Leasing Companies

While not entirely nontraditional, companies such as TelMark and FCS Leasing Corp. have posted impressive recent growth and have generated more attention with creative programs and products. They are included at this point because the roles they will likely serve can be understood in the context of state-space models and the general evolution of financial markets. Analogies from the automobile and airline industries are a bit more difficult to draw than in the case of captives, but the current posture is still for continued aggressive future growth.

In more nearly matching or smoothing the rate of physical depletion of capital with that of the associated financial flows, leasing provides a direct "spanning" of a dimension of cash-flow variability that can be very attractive. This benefit may accrue explicitly as part of a sophisticated cash management system or implicitly as the result of the direct implementation of leasing terms. In some cases, there remains a potential tax advantage to leasing, and leasing can assist in managing the composition of the balance sheet.

As with captives and vendor finance units, it is not clear whether traditional lenders view leasing markets as friend or foe. Many of the larger, more successful leasing efforts are outgrowths of traditional lenders or captives that evolved from lending to leasing. Nonetheless, leasing substitutes for other forms of debt capital that the lender might have provided, and can be appropriate when the structure of the borrower makes other forms of debt inappropriate or undesired.

Securitization, Swapping, and the Separation of Origination From Warehousing

The natural maturation of financial markets promotes more finely separated functions of lending. Securitization can fund borrowers at competitive rates by pooling efficiencies and matching assets with holders who can utilize their properties most efficiently. In large part, traditional lenders already enjoy many of these benefits, but smaller and newer lenders may be able to fund relatively specialized programs through this means.

From the user's perspective, securitization both manages the size and composition of the balance sheet and provides liquidity. The appropriate role depends on the institutional circumstance and the current competitive conditions. For securitization to become more important, though, it must be demonstrated to be of additional value to the capital markets and not already available through other avenues. Traditional lenders may discover that they can thwart the development of upstart securitizing operations by providing those services directly. Several types of securitizing operations are likely to continue to compete for market shares of funding—including long-term assets via institutions like Farmer Mac, as well as selling short-term receivables by factoring or managing the size of the balance sheet through purchase, sale, or swaps of revenue flows.

But why have agricultural securitization markets lagged many others? Is this due to some uniqueness in agriculture or the difficulty in standardizing claims, or have the relatively sound aggregate capital structure and well-developed alternative funding mechanisms in agriculture simply prevented the need? Firms such as Deere, First Chicago, and AgServices have demonstrated great facility with securitization of agricultural claims, yet in aggregate the incidence is fairly sparse. Whatever the reasons for the current state, the potential benefits of securitization will be more readily available to agriculture in the future.

Investment Banking, Quasi-Equity, and Development
of New Equity Markets

Historically, equity markets are depended upon to facilitate the maturation of an industry, and then debt markets can accrete to larger relative shares. In agricultural production, there seems to be a paradox in that the industry has relatively high aggregate equity ratios, but individuals with debt are frequently very highly levered. Why do such great differences in equity ratios exist across otherwise similar units? And how well do current market products facilitate the removal and infusions of capital from production units? Agriculture's continued commercialization (increased coordination, larger units, etc.) is increasing the need for equity products to capitalize production. The innovators that provide this equity will control an increasing share of the production sector. Equity providers may remain largely individuals, or quasi-equity products from integrators, guarantors, input suppliers, and venture capitalists may continue to develop. In any case, the traditionals again need to determine whether to be "deal makers" and assemblers of capital or to passively provider debt capital only. If agriculture continues its structural evolution to larger, more commercialized units, deal makers will be more in demand than debt brokers.

Participation of Traditionals in Nontraditional Markets or Products

Traditional lenders, though beset by change, are well-positioned to fund the emerging needs of their customers. Their comparative advantages in many of the functions of intermediation can be exploited while transferring out the functions that are better performed elsewhere. For example, the Trade Credit program of the FCS is designed to utilize vendors' close contact with customers without forcing vendors to acquire the skills or resources to make credit decisions or secure funding. Traditionals are also moving into related businesses and have established partnerships with other providers of financial services in cases where direct supply is prohibited by regulation. Examples include affiliations with securities and investment houses by banks; or housing of insurance providers, IDS services, and so on in Farm Credit offices.

The future of agricultural lending will be more customer-driven, but who will be the customer? It may be a re-lender, end user, related product vendor, equity provider, and so on. But in any case, successful financial products will need to be tailored to meet specific customer needs.

Conclusion

The cruel efficiency of financial markets serves as a basic tenet of economics. Thus, the basic valuation question related to new lenders' products will be revealed through time. Some informed judgments, however, are reasonable in light of the evidence, recent trends, and through analogy to other experiences. Among those:

- Financial markets are continuing to provide more specialized capital products for more specialized needs.
- The growth in nontraditional delivery points will continue.
- Product line offerings at both traditional and nontraditional lenders will proliferate.
- Identities of players will give way to functions performed as a means of competitive distinction.
- Changes in industrial organization of lending will parallel continued commercialization of agriculture.
- More interlinkages between traditional and nontraditional lenders will be formed.

While the generalities are easy to guess, aggregate impacts on the quantity of credit supplied and demanded are, of course, extremely difficult to quantify. However, at least two scenarios are plausible. First, if nontraditional lenders do have a true cost advantage in supplying debt capital, it could lead to a substitution in the firm's financial structure away from equity toward the lower cost debt. Likewise, further maturation of the agricultural industry would be consistent with higher aggregate debt levels. Alternatively, vertical linkages and complex ownership arrangements could decrease the aggregate demand for debt because of the use of equity capital and quasi-equity products in more coordinated production units.

Equally difficult to discern is the future policy environment. Historically, credit programs promoted the desired organization of the farm sector, and were used to protect favored sectors of the economy. The trend, however, is now to limit direct government involvement in favor of guarantee programs and subsidies administered indirectly. At the same time, the tradeoff between access to new authorities and the acceptance of financial institutions as instruments of public policy argues that the indirect effects of government influence in the financial sector may increase. To the extent that the most efficient output-producing organization of the sector is not the objective of policy, private market alternatives to promote those types of policies are unlikely to arise. Hence, young farmer programs, special production programs, and other historically favored programs are not likely to be as great a part of the future financial landscape. Other policy changes, such as a reduction in capital gains tax rates, could prompt some equity retirements from agricultural holdings that could be substituted with debt by new owners.

Finally, as the agricultural sector continues to lose its distinction from other sectors, its access to financial products and innovations already available to other sectors will improve. The evolution of financial products and the resulting capital structure may not be as profound as in the automobile, airline, or consumer products industries, but emerging lenders and their products will play an ever increasing role in financing agriculture.

Notes

1. Although vendor finance operations have been in existence for several decades, their significant increases in volume and the systematic expansion of their activities permits their inclusion in this category.

2. Deere has been able and willing to finance non-Deere inputs for a long time, although usually at dealerships with an affiliation to Deere. The recently publicized agreements to fund and perform the credit evaluation functions for vendor finance operations of Pioneer AgriGreen and Growmark's AgriFinance program (previously funded by CoBank), as well as their credit card businesses represent further extensions of their activities into territories previously occupied by traditional lenders.

3. Except in cases where all but two of the input variables are perfectly correlated, in which case it could be argued that there are really only two risk dimensions. Then, two output claim dimensions may span these risks.

References

Dipchand, Cecil, Gordon Roberts, and Jerry Viscione. 1982. "Agency Costs and Captive Finance Subsidiaries in Canada." *Journal of Financial Research* 2 (Summer), 189-199.

Dodson, C. 1996. "The Changing Structure of Nonreal Estate Markets", in *Agricultural Income and Finance: Situation and Outlook Report*, AIS-60, U S Dept. Agr, Econ. Res. Serv., Feb.

Fooladi, Iraj, G.S. Roberts, and J.A. Viscione. 1986. "Captive Finance Subsidiaries: Overview and Synthesis," *The Financial Review*, 21 (2) (May), 259-275.

Henricksen, W., and M. Boehlje. 1995. "Captive Finance Companies: Are they Cost Competitive?" *Journal of Agricultural Lending*, 9, #2 (Fall), 25-28.

Lewellen, Wilbur G. 1972. "Finance Subsidiaries and Corporate Borrowing Capacity." *Financial Management* (Spring), 521-537.

Remolona, Eli M., and Kurt C. Wulfekuhler. 1992. "Finance Companies, Bank Competition, and Niche Markets." *Federal Reserve Bank of New York Quarterly Review* (Summer), 25-38.

Roberts, Gordon S., and Jerry A. Viscione. 1981a. "Captive Finance Subsidiaries and the M-Form Hypothesis." *Bell Journal of Economics* (Spring), 285-95.

Roberts, Gordon S., and Jerry A. Viscione. 1981b. "Captive Finance Subsidiaries: The Manager's View." *Financial Management* (Spring), 36-42.

Sherrick, B.J., S.T. Sonka, and J.D. Monke. 19??. "Strategic Change and Competition in the Agricultural Credit Market", *Agribusiness: An International Journal*, 10 (4): 1-17.

Sherrick, B.J., and R.W. Lubben. 1994. "Economic Motivations for Vender Finance", *Agricultural Finance Review*, 54, pp. 120-131.

Staten, Michael E., Otis W. Gilley, and John Umbeck. 1990. "Information Costs and the Organization of Credit Markets: A Theory of Indirect Lending." *Economic Inquiry* 28 (July), 508-529.

U.S. Department of Agriculture, Economic Research Service. 1997. *Agricultural Income and Finance: Situation and Outlook Report*, AIS-64, Feb.

PART FOUR

POLICY IMPLICATIONS

8

Public and Private Policy Implications

Cole R. Gustafson, Marvin Duncan, and Jerome M. Stam

An industrial realignment is occurring in the agricultural and food system as the United States moves into the twenty-first century. Industrialization is an expression of natural market evolution, and not a ruthless attack on the family farm structure by exporters, processors, and input suppliers. Producers, input suppliers, and now processors are establishing business alliances and taking ownership across two or more of these three components of the U.S. agricultural sector. Economic rewards are being given to producers who commit to deliver consistent, quality products, and who pursue economies of scale. In doing so, consumers' wants and needs are expressed back into the production and distribution system. Such an arrangement provides consumers with the optimal quality, availability, and price.

In a new era of increased risk, producers willing to enter integrating relationships will find themselves more profitable than their independent counterparts. Furthermore, industrialization demands considerable investment in technology, which in turn typically requires substantial amounts of innovative credit and other new forms of financing. Farms that have lost prominence typically are those that quit investing in technology.

These realities do not bode well for many independent family farms or ranches. The transition from marketing commodities to marketing products will increasingly require integration and coordination. Larger farm operations will be in the best position to take advantage of changing market conditions and innovative financing arrangements.

Of course, a description of farming (with regard to financing the evolving structure of U.S. agriculture) requires more information than simply farm numbers, sizes, and products (Sommer et al.1997). As farm numbers have declined and farming has become more competitive in both domestic and global markets, the business of operating a farm has become more complex. Technological advances alone induce farmers to continually modify business decisions about input use, product mix, and production practices.

However, other management decisions also have become critical for farm business survival. For example, farm operators may use contracting to share the risks of production or marketing, participate in government programs—if available—to protect against market fluctuations, purchase available crop or risk insurance, utilize hedging and other related tools, or lease rather than buy land and equipment in order to avoid indebtedness. The differing characteristics of operators and their households are critical to understanding how farm businesses differ and how they use financing.

It is important to place the changes in agriculture and its financing in context as the United States moves into the next century. The private and public mix regarding the financing of the agricultural sector has evolved in a complex manner since the first part of the century, with roots extending back much earlier. Government consciously employed various forms of credit institutions and programs to facilitate the transformation of the farm sector and related subsectors in an effort to enhance productivity and incomes. But, while the issue of rural credit is an old one, the direct federal role only extends back just over eight decades. It is important to briefly examine the public-private farm and rural credit record to better place the current structural and financing revolution in context.

Private and Public Farm Credit

The historic rationale for public intervention in credit markets is usually twofold: (1) the concept that private credit markets cannot meet social objectives and priorities affecting the allocation of resources and distribution of income, and (2) the perceived imperfections in private credit markets that result in credit rationing, market failure, and other types of credit gaps. A more recent concern in the literature is the problem of asymmetric information (Barry 1995). Market imperfections constrain the effectiveness of private credit markets and therefore reduce efficiency and overall welfare attainment (Barry 1995). However, public and quasi-public credit intervention to correct private market imperfections typically is expensive.

Public credit activities historically have been predicated on filling alleged credit gaps. A nominal credit gap exists when the financial intermediation—the process of channeling funds from savers to borrowers—does not allocate loanable funds efficiently and to the highest return uses.[1] Credit gaps occur when artificial barriers block the flow of funds to sound investment opportunities whose benefits (returns) exceed their cost (interest rate). Artificial barriers may result from legal and institutional restrictions, inadequate information, prejudice, habit, and tradition (McKie 1963).

The public may benefit when the government takes action to correct a credit gap or other competitive imperfections, but direct government lending and the activities of government- sponsored agencies divert funds from other sectors of the economy. Even when the public entity guarantees private investment, it causes private debt to be partly interchangeable with public debt. This alters the supply-demand balances in the different sectors of the capital market and, as a result, interest rates and the allocation of resources in the economy (McKie 1963). If credit gaps distort the flow of loanable funds from socially optimal patterns, however, correction of the distortion improves the allocation of resources.

Public credit programs and financial policies have been a significant factor in agricultural finance via government loan programs, government sponsored enterprises, and depository institution regulations. But such federal government activities have only existed in a significant way in recent decades. It may be difficult, given the major role they have played, to realize that there was a time in U.S. economic history when no federal government rural credit activities existed.

The Quest for Credit Access

U.S. farmers' access to credit was an issue even in colonial times. In 1932, T. N. Carver noted, "During the whole of our colonial period and the first century of our national life...the problem of financing the farmer was one of our major economic problems. It even played a larger part in politics than any other questions except those of slavery and the tariff" (Sparks 1932, p. xi). Another observer lamented that much of the literature on farm credit covered only the period since the establishment of the Federal Farm Loan system in 1916 and that "a popular impression prevails that very little had been done toward meeting the credit needs of American agriculture before this recent legislation" (Sparks 1932, p. vii).

During the last part of the nineteenth and into the twentieth century, farm sector demand for financing was modest, given the simpler type of agriculture and lower capital requirements than the capital-intensive farm sector that was to follow. The majority of inputs used were home-produced. Seeds were saved from a prior crop for planting, fertilizers consisted primarily of animal manures, and weed control was accomplished by hand and/or animal labor. Mechanization was in its infancy. The few tools and machines needed for production were primarily purchased with residual cash saved from periods of bountiful production. Commercial fertilizers, insecticides, feed supplements, and fuel had not yet become typical farm operating expenses. Even methods of transferring land from generation to generation (e.g., gifting and exchanges of labor for capital) minimized capital demands.

A difficult farm credit situation persisted for decades, particularly with respect to real estate loans (Brake 1974, Cochrane 1979). In the late 1800s and early 1900s, such farm credit was difficult to obtain on reasonable terms, and sometimes impossible to obtain on any terms. Prevailing interest rates ranged from 7 to 12 percent, with the farm mortgage company often assessing an additional 1-2 percent for the agent's commission. Loans were for 3-5 years with the farmer expected to repay the entire principal upon loan maturity. Renewal was often possible for another short period, depending in large part on local economic conditions. But economic circumstances sometimes forced banks to call loans for full repayment and, if the farmer could not locate another credit source, foreclosure often resulted.

Thus, extending back well into the latter part of the nineteenth century, the problem of "rural credits" was on the political agenda and a topic of great debate, but concrete results stemming from the Populist and other movements took a long time (Nourse 1916). The depth of the early antipathy toward federal assistance to agriculture is difficult to fathom today given the public record of assistance to come in subsequent decades. For example, agricultural drought relief was not viewed as a federal responsibility in the nineteenth century. In 1887, Congress appropriated $10,000 to provide seeds for drought-stricken Texas, but President Cleveland vetoed the bill, arguing that the government has neither the power nor the duty to relieve suffering and that paternalism weakens character. People, he stated, support the government, but the government should not support the people (Dyson 1988).

Farming was the economic engine driving rural America until after 1950, when other rural sectors began to predominate. Thus, policies strengthening a problem farm sector would promote economic progress for all rural America. The persistent rural problems caused President Theodore Roosevelt in 1908 to appoint a Country Life Commission to study all aspects of rural living. One of the most prominent rural deficiencies identified was the "lack of any adequate system of agricultural credit, whereby a farmer may readily secure loans on fair terms..."(U.S. Country Life Commission 1909, p. 15). In 1910, 32.1 million people lived on 6.4 million farms and comprised 34.9 percent of the U.S. population.

The rural credits debate focused on two subtopics: the level and diverse pattern of interest rates offered to farm borrowers in various sections of the country, and conditions of farm tenancy. The existence of nonuniform interest rates on farm mortgages suggested to many that there were inadequate credit facilities for agricultural lending. These inadequate facilities were also blamed for the large number of tenant farmers. It was thought that because capital markets were imperfect, tenants lacked access to credit facilities and thus were unable to purchase land of their own. By 1912, rural credits had become a major issue with planks addressing it a part of all three major political party platforms (Cochrane 1979).

Because commercial banks were historically enjoined from farm mortgage lending, farm finance was largely provided by farm mortgage brokers and by life insurance companies (O'Hara 1983, p. 426). The farm mortgage brokers financed their lending by issuing bonds, while the life insurance companies relied upon premiums to fund their mortgage commitments. In 1913, the Federal Reserve Act (P.L. 63-43) loosened restrictions on commercial bank agricultural lending and, as a further concession, installed a farmer as one of the system's directors. But advocates of increased agricultural lending were not satisfied with rural credit availability, alleging that the commercial banking system was not meeting the needs of farmers given that the system's many small units favored shorter-term loans and generally tried to avoid much of the risk faced by the farm sector. A new intermediary designed solely to provide rural credits was demanded. The leaders of this movement looked to Europe for a rural credit model (Nourse 1916).

In 1912, President Taft requested that U.S. ambassadors serving in Europe investigate and report on the cooperative rural credit systems in the countries to which they were assigned (Brake 1974). That same year a private group—the Southern Commercial Congress—appointed the American Commission to visit Europe and report on cooperative credit systems there. In 1913 President Wilson appointed a U.S. Commission of seven persons to study European rural credit systems. The two commissions made a joint study in 1913 that was presented as a preliminary report to Congress (Hoag 1976). In 1914, the two groups made more extensive and separate reports to Congress. The reports prompted three different proposals for new agricultural credit institutions (Brake 1974): (1) to obtain funds for a new credit entity through the sale of mortgage bonds to investors, (2) to organize cooperatives to serve farm lending needs, and (3) to provide direct government loans to farmers.

The Federal Government Acts

Federal farm credit action began with the Federal Farm Loan Act (P.L. 64-158), signed by President Wilson on July 17, 1916. It provided both for federal land banks (FLBs) with borrower participation and for joint-stock banks to be formed and owned by private investors (Sparks 1932). The establishment of a dual system of farm mortgage lenders in 1916 was the result of an awkward political compromise setting up the FLBs side by side with the joint-stock banks (O'Hara 1983). Both FLBs and joint-stock banks were authorized to issue tax-exempt bonds like the private-purpose municipal bonds of today. The Federal Farm Loan Act envisioned a rural credit system with two separate systems providing the same service (Sparks 1932).[2]

The result of the initial action in 1916 was the formation of the very first government-sponsored enterprise (GSE) in the form of FLBs. The quasi-governmental FLBs were created to make long-term loans on farm real estate. In 1923, the Federal Intermediate Credit Banks (FICBs) were formed to discount short-term agricultural loans made by other farm lenders. In 1933, Production Credit Associations (PCAs) were established to make short-term operating loans to farmers, and Banks for Cooperatives (BCs) were set up to finance purchasing and marketing cooperatives. FICBs were the financial intermediaries that provided credit to Production Credit Associations (PCAs), while at the same time continuing their post-1923 task of lending to other non-PCA financial institutions, such as commercial banks. Organized into 12 districts for each type of bank, the entire complex by 1933 became the Farm Credit System (FCS). It has undergone continual changes in authority and structure pursuant to major legislation in 1953, 1971, 1980, 1985, 1986, and 1987 (Brake 1974, Collender 1992, Collender and Erickson 1996, Lee and Irwin 1996). A key feature of the FCS as a rural-focused GSE is that it obtains its funds on the national money market slightly above U.S. Treasury rates, making it a formidable competitor at the retail loanmaking level for private lenders such as commercial banks.

Direct credit assistance for farmers was also on the political agenda from the early 1900s, with a number of advocates and detractors (Nourse 1916, pp. 772-776). Federal lending to farmers has an 80-year history, with the emphasis in the early years on *disaster loans*. The disaster loan concept was introduced during World War I. President Wilson, in a letter dated July 26, 1918, authorized $5 million in loan aid to farmers in several drought-stricken areas. This was the first time that the federal government became involved in making loans to farmers. From 1918 to 1937, various authorizations were made for issuing emergency crop and feed loans for a limited period of time, generally 1 year.

Although the U.S. farm sector began to experience financial problems in the 1920s, the cumulative economic problems of the Great Depression of the 1930s were the first massive negative financial shock this century. The severe financial hardship of the farm family gave impetus to the continued growth of federal direct lending programs for farmers (USDA 1989, USDA 1991). The Hoover administration (1929-33) saw a great expansion in federal expenditures for drought relief (Dyson 1988). In President Hoover's last year, there were 508,000 loans totaling $64 million for feed, seed, fertilizer, and fuel. This expansion continued under the Roosevelt administration. Supervised, low-cost loans and grants became available in 1933 to help farm families remain on their farms or re-establish themselves in farming. Farmers made up 24 percent of the population in 1935, while farm numbers were at their historical peak of 6.8 million.

Congress transferred these programs to the Resettlement Administration in 1935 and granted the agency authority to make supervised farm real estate loans to tenant farmers with few financial resources. The Bankhead-Jones Farm Tenant Act of 1937 (P.L. 75-210) gave the Resettlement Administration authority to provide long-term financing to farmers lacking other sources of credit in order to purchase land and improve their farms and homes. The Resettlement Administration was the beginning of the "official" lineage—now extending back 63 years—of financial and technical assistance to farmers through what later became the Farmers Home Administration (FmHA) and the Farm Service Agency (FSA) (USDA 1989). In 1937, the Resettlement Administration was transferred to USDA and renamed the Farm Security Administration. The Farm Security Administration at that time was to make farm rehabilitation and farm ownership loans to farmers unable to borrow from usual sources of credit. The

Farmers Home Administration Act of 1946 (P.L. 79-731), which took effect in 1947, established FmHA. Following the signing of the Federal Crop Insurance Reform and Department of Agriculture Reorganization Act of 1994 (P.L. 103-354), the FmHA was abolished and its farm credit programs were transferred to the newly created Consolidated Farm Service Agency (CFSA) effective October 13, 1994. CFSA was subsequently renamed the FSA in 1995 and currently operates out of nearly 2,400 offices administering USDA's farm credit and other farm assistance programs.

FSA (and its predecessor FmHA) is a public "lender of the last resort" in a safety-net role to farm borrowers unable to acquire commercial credit at reasonable rates and terms. The characteristics of such USDA farm borrowers have evolved through time. Historically, these generally have included impoverished or destitute farm families, young farmers entering the sector, small yet potentially viable farms, and larger farms experiencing significant distress due to natural disasters and economic emergencies (Barry and Boehlje 1986). These types of farmers do not have access to commercial credit, but are considered candidates for public credit programs. FSA (and its predecessors) serves as a source of supervised credit and technical support in the hope that such assistance will improve the financial viability of farming enterprises enough for them to qualify for private-sector resources.

FmHA-FSA policies and legislation place a high priority on keeping struggling farmers on their farms. FmHA was initially viewed as a social welfare agency as much as a credit agency. But FmHA was much less a welfare agency than its predecessors because Congress diminished its welfare orientation. The 1946 FmHA Act restructured USDA's credit mission, dropped farm rehabilitation programs, and focused on farm operating and farm ownership loans. The move away from FmHA's welfare function continued to evolve through time. For example, the Rural Development Act of 1972 (P.L. 92-419) authorized the making of FmHA-guaranteed loans by commercial lenders for farming. Historically, the majority of FmHA's and FSA's annual lending came through their direct programs. As a result of new policies in the 1980s, about 80 percent of the $2.86 billion in annual farm lending authority now comes through FSA's guaranteed loan programs. Outstanding FSA direct and guaranteed farm loans stood at $16.3 billion at the end of fiscal 1997.

The Credit Gap Turns Into Credit Abundance

In retrospect, it can be argued that U.S. agriculture was not only severely depressed economically between World War I and World War II, but was undercapitalized as well (USDA 1984). The federal government responded by introducing programs to reduce the riskiness of farming and to ensure easier access to credit at more favorable terms. The FCS and FSA's predecessor agencies (along with USDA's Commodity Credit Corporation) dramatically changed the credit situation and flow of funds and resources to agriculture that facilitated the financing of a technological revolution and capital restructuring of U.S. agriculture between the 1930s and 1970s. They were so effective that by the 1960s, the evidence from several research studies indicated that excess agricultural capacity was likely (USDA 1984). During the 1960s into the early 1980s, funds were plentiful relative to national demand with farmers having easier access to loans and at lower interest rates than did others in the economy.[3] This situation helped many farmers to prosper and to accumulate assets ensuring a productive U.S. farm sector, but it also may have contributed to inflated prices of farmland and other capital goods.

The 1970s were generally a good time for agriculture, with optimistic expectations over worldwide demand for U.S. farm products and inflation. Agricultural exports expanded as the dollar declined in value. Prices for farm commodities rose early in the decade in response to strong demand for feed grains and wheat. Farm production and investment expanded in a climate of low, and at times negative, real interest rates. In this economic boom, farm borrowing grew and land values increased rapidly. Lenders, consultants, and others often encouraged additional leveraging via borrowing to finance expansion, some of it speculative. Rising machinery investment levels combined with land price and other cost increases resulted from the boom times and ample credit.

The early 1980s saw a rapid turnaround in the forces that had caused the expansion. Back-to-back recessions in 1980 and 1981-82 hit the farm sector hard. A large increase in the value of the dollar reduced the demand for U.S. farm exports. Other countries expanded agricultural production in response to generally higher world prices. In the United States, the cost of producing commodities increased into the early 1980s. Monetary policies designed to reduce inflation prompted interest rates to rise to unprecedented levels in the early 1980s. Farm input costs increased, while net farm income generally fell. Returns to land declined due to a reduction in exports and commodity prices, a high cost structure, and even lower projected returns. The declining farmland values weakened farmers' equity positions and their collateral values used to back loans. Some farmers were unable to make principal and interest payments on the large amount of debt acquired during the 1970s boom period. These interrelated economic changes in the 1980s caused the most severe financial stress for the farm sector since the Great Depression of the 1930s.[4] This occurred when there were 2.4 million farms with a much greater capital intensity and a much differently structured rural sector than existed 50 years earlier.

Deregulation also became an important reality as moves were made to make bank regulation consistent with an efficient and competitive banking system (Spong 1990). In the early 1980s, considerable regulatory and other changes in the U.S. financial markets affected the agricultural sector. The Depository Institutions Deregulation and Monetary Control Act of 1980 (P.L. 96-221) and the Garn-St. Germain Act of 1982 (P.L. 97-320) substantially deregulated commercial banking. Both geographic and product-line barriers that had existed for a long time in the financial services industry were significantly reduced. These changes caused forced rural areas to compete for funds in a national market. The Farm Credit Act Amendments of 1980 (P.L. 96-592) proposed to update and improve the operation of the FCS. The deregulation, coupled with changes in monetary policy and fluctuating inflation rates, significantly altered the financial environment in which agricultural lenders and borrowers were required to function. The number of commercial banks declined from 14,434 in 1980 to 9,215 at the end of the third quarter of 1997.

The financial hardship experienced by farmers in the 1980s and indirectly throughout rural areas spurred the federal government to undertake specific credit initiatives to assist with economic adjustment. Special credit programs implemented to assist financially stressed farmers and their lenders helped preserve the operation of many farms that would have been forced out of business. That assistance, in turn, affected farm-service industries and rural communities.

By the early 1980s, the farm sector had accumulated more debt than could be repaid from current and expected future income. The national farm business debt, which peaked in 1983 at $193.8 billion, had accumulated since the mid-1970s as farmers expanded production to meet anticipated future domestic and export needs. Some farmers speculated that farmland values would continue their rapid appreciation. The expansion was further fueled by an inflating economy and negative real interest rates.

Government farm policies also often encouraged the expansion. For example, the federal government greatly expanded capital to the sector through the FmHA. FmHA increased annual farm lending from $1 billion in 1974 to over $8 billion by 1981. FmHA supplied a total of $34 billion in farm credit from 1975 through 1981. The increase raised FmHA's share of total agriculture business debt from 5.4 to 15.7 percent. The quasi-government FCS and the FmHA together held 35.1 percent of farm business debt in 1975 and this increased to a peak of 46 percent in 1981; it was 30.9 percent in 1997.

The federal government responded to the farm financial difficulties of the 1980s with a range of policies to provide farmers with income support, credit assistance, and new legal rights as borrowers (Stam et al. 1991). During the 1980s, net Commodity Credit Corporation outlays for commodity programs reached record highs, totaling $133 billion. USDA's Economic Research Service calculated that federal credit subsidies to agriculture during 1986-88 alone totaled $7.2 billion. The financial stress caused considerable retrenchment and restructuring among the farm lenders (Collender 1991, Collender 1992, Collender and Erickson 1996, Peoples et al. 1992, Stam et al. 1991, Stam et al. 1995). During 1984-90, commercial banks, FmHA, and FCS charged off $18.6 billion in farm loans.[5] Congress authorized $4 billion in Treasury- guaranteed bonds for a FCS bailout, of which $1.26 billion was issued to aid FCS restructuring.

The Agricultural Credit Act of 1987 (P.L. 100-233) included a provision to create the Federal Agricultural Mortgage Corporation (Farmer Mac) to operate a secondary market for farm mortgages in the private sector and for a separate market for FmHA-FSA agricultural loan guarantees (Hiemstra et al. 1988). The purposes of the new private market were: (1) to increase availability of long-term credit to farmers at stable interest rates, (2) to provide greater liquidity and lending capacity in extending credit to farmers, and (3) to facilitate capital market investments in providing long-term agricultural lending (including funds at fixed interest rates) (Barry 1995). The FmHA-FSA secondary market allowed lenders to sell the guaranteed portion of their loans to investors, recouping the loan principal but retaining servicing responsibilities.

A Mature Farm Credit Delivery System

The public-private credit delivery system that serves rural America has grown to be very complex (USDA 1997c). Federal quasi-public (FCS) and public (FSA) institutions historically have dramatically changed the credit situation and flow of funds to agriculture.[6] Credit availability has facilitated and encouraged structural change in the farm production sector (Barry 1995, Lins 1979). During the 1960s to the early 1980s, farmers had easier access to loans and at lower interest rates than did other sectors of the economy. The scarcity of credit noted in the first half of this century had turned into abundance. Yet, in retrospect, there probably was a credit gap facing agriculture prior to the development of federal programs in the first half of this century. Agriculture was undercapitalized until World War II because of inadequate financial markets. The federal government increased the supply of credit by creating special institutions and programs serving agriculture. This infusion of credit helped finance the technology revolution of U.S. agriculture.

Earlier farm problems, such as poverty and tenancy, have been largely resolved. Farm incomes and returns are comparable to those found in nonfarm sectors (Holmes et al. 1991, USDA 1997a). Private lenders have been innovative in serving agriculture and better recognize the unique requirements of lending to agriculture. The legal, institutional, and economic framework encourages a diversity of credit resources for farmers. Financial markets have been significantly deregulated, the FCS continues to restructure, FSA has downsized and emphasizes guaranteed loans, and a secondary market for farm mortgage loans exists through Farmer Mac. The result is that farmers are no longer seriously disadvantaged in their access to credit services relative to borrowers in other business sectors.

Farm lenders, since the credit crisis ended in the mid-1980s, have emphasized working with larger loans, thus moving up the size scale of farm businesses (Stam 1995). Thus, as Collender and Koenig ask in chapter 6, do substantial public benefits accrue from a quasi-public institution, such as the FCS, lending to a core of commercial farmers while paying little or no attention to public-purpose issues such as loans to beginning, younger, or limited-resource farmers?[7] Moreover, the FSA loan programs occupy a niche lender role compared with the massive attention paid to farm poverty via supervised lending in an earlier era by the former FmHA.

As the agricultural sector entered the 1990s, several important changes transformed the ownership patterns of assets; the scale and size of farming units; the independence of suppliers, producers, and processors; and the adoption of technology in production processes. Combined, these shifts created a demand for an entirely new portfolio of financial services and products. The agricultural lending industry has responded only modestly to this demand, primarily through refinement of its traditional products and services. However, given the depth and rapid pace of change occurring in the agricultural sector, a more fundamental restructuring of the agricultural lending industry is required and underway.

Conditions are radically different as we enter the twenty-first century. Going into this century, the credit problems facing the farm sector have changed and have been significantly reduced. The problem now is financing a large-scale, high-tech farm sector (Harrington et al. 1995, Moss et al. 1997). For example, in 1996 some 336,400 farms with incomes exceeding $100,000 per year accounted for 16.3 percent of all farms and 81.5 percent of all cash receipts from farm marketings (USDA 1997b). Some 49,500 farms (2.4 percent of the total) were responsible for 43.7 percent of all sales.

Structural Change in Sector
Requires New Lending Paradigm

The rapid horizontal and vertical integration of production units that is occurring in the farm sector during the 1990s, as described by Harrington and others in chapter 2 and Drabenstott and Barkema in chapter 3, will require significant changes in the underwriting and credit administration procedures of both public and private firms extending debt capital to the agricultural sector. The transition from financing a large number of medium-sized, local, sole-proprietor family farms requiring mostly debt capital for real estate financing to financing a small number of large, geographically dispersed, and complex firms that primarily require working capital will be difficult.

Moreover, financing needs for more traditionally organized farms will continue to be an important focus for most lenders. That traditional focus will not, however, represent the primary opportunities for growth and credit innovation. Those will occur to better service the large integrated farming businesses. Some of the innovations may later become part of a lender's product line to its traditional customers.

Large integrated farming businesses possess four distinctive characteristics. First, they are not homogeneous with respect to commodities produced, methods of production, and financial organization. Second, they are heavily dependent on information-based technology that is difficult for an outsider to acquire, readily observe, and evaluate. Third, the scope of their operation typically is much broader than for a traditional farm business. Input supply, production, and marketing/processing functions—or important parts thereof—are coordinated by a single management entity, as opposed to three or more individual firms. Finally, their capital structure likely is different. Assets are more often controlled through lease and rental arrangements than through ownership. Long-term capital requirements may be financed through sales of equity instruments such as stock. Ongoing business financing needs of these emerging firms are primarily cash-flow rather than asset-based. And, these firms may require access to a broad array of other financial services such as cash management, risk management, payroll management, inventory financing, receivables management, foreign trade financing, debt placement, and merger/acquisition.

The credit underwriting and administrative procedures followed by public and private lenders in the past century were well suited to the debt financing needs of the sector. In most geographic areas, firms consisted of a large number of homogeneous production units. Examples include the typical 600-acre corn/soybean farm of the Corn Belt, 320-acre dairy farm of the Lake States, and the 1,500-acre wheat/beef farm of the Northern Plains. Productivity of units in each region was highly related to the number, composition, and management of physical assets operated (i.e., more acres/head of livestock generated increased farm receipts). The slow evolution of these firms over time allowed lenders to rely on intuitive, informal benchmarks and "rules-of-thumb" in their credit decision-making process. Most lenders had farm backgrounds and could easily evaluate a borrower's use of debt proceeds when driving by their residence and conducting a "windshield survey" of asset utilization. Moreover, since the lender was a resident of the community in which the loan customer resided, informal communication networks provided valuable information about the production practices and business acumen of the borrowers

Since production practices and technology were relatively uniform within a geographic area, any deviation in crop performance from field to field was attributed primarily to management. Tardy spraying of weeds resulted in weed-infested stands, lower yields, reduced farm receipts and incomes, and eventually declining credit availability. Asset-based lending policies were particularly well suited to this credit environment because of their low administrative costs, relative lack of risk for the lender, and ease of implementation.

Evaluating the creditworthiness and business performance of horizontally and vertically integrated firms is far more challenging. The heterogeneity and unique combinations of assets embodied in these firms implies unique credit underwriting standards for each unit. The agricultural lending officer's information base will be strikingly different from the past. Whereas technical knowledge of farm/ranch operations was a prerequisite for past success, the successful loan officer of the future will need to identify and evaluate the economic value of firm alliances, information sources, and coordination methods. The economic assets of highest value in these firms are the strategic alliances and business relationships among firms. As Boehlje implies in chapter 4, the ownership interests, liability, and methods of legal recourse underpinning these alliances and relationships are difficult for not only creditors but also owners to identify. These ownership interests are widely scattered geographically, complicating the ability of a single loan officer to monitor each account. Moreover, the ownership interests typically are involved in a number of disparate business lines that often extend well beyond production agriculture, which again pressures a traditional loan officer who may have narrow experience in a limited number of agricultural enterprises. Finally, the expertise of traditional loan officers primarily was in commodity production and marketing, not production, processing, and marketing of differentiated, branded products—the hallmark of integrated agricultural firms.

Whereas the creditworthiness of traditional farms was assessed according to asset levels and utilization, the creditworthiness of integrated firms is based on information-based technology, cost-effective asset control, and effective execution of complex business relationships. Proprietary technology and methods of production, sophisticated coordination of management information across units, and relationship marketing all combine to determine firm financial performance. Information-based resources are not readily observable, difficult to value, and vary considerably from firm to firm. Similarly, evaluation of business management skills is difficult, yet crucial to the creditmaking decision. Consequently, creditworthiness of today's integrated firms is very difficult to appraise.

Finally, the sheer scope and size of integrated firms will probably tax the lending capacity of local credit institutions. Lending limits constrained by lender capitalization will mean that for most community-based lenders, it will be necessary to develop strong correspondent linkages to handle credit over lines or to bring together a consortium of lenders to service the credit line. Indeed, the larger farm businesses likely will insist on doing business with a group of lenders working with the borrower through a lead bank arrangement.

Local credit institutions also will have great difficultly providing the range of financial services demanded by integrated agricultural firms. The ad hoc structure of existing correspondent relationships does not yield the permanency, level of resources, and breadth of financial services required by these large and complex firms. Lenders will need to make considerable investments in employee education, development of new lending policies, and business procedures if they are to keep pace with the financial services requirements of these large businesses.

In deciding whether or not to serve this emerging cadre of integrated farm firms, lenders will want to weigh the costs of doing so against the benefits to be achieved. Here, two different strategies emerge. One is to view each transaction on its own merits and its profitability to the lender. This strategy works best for a mortgage loan or an equipment loan. A more compelling strategy, however, is to evaluate the potential for a long-term relationship with a borrower and to ask whether, over the length of that relationship, investment in new capacity by the lender will pay off in future profits. In short, what is the customer's potential in terms of service demand? What will it cost the lender to gear up to serve those needs? Will the present value of the future income generated with that customer, or class of customers, exceed the present value of the associated costs? Will doing so maximize the present value of the net revenue stream for the lender, or are there other strategies that will do so? Thus, it should not be surprising that structural changes on the part of lenders' customers will result in greater segmentation of the agricultural lending market.

Market Segmentation and Targeting by Lenders Uncertain

Harrington and others in chapter 2 present a useful delineation of farm types that are likely to emerge in the next decade (Figure 2.4). Likewise, Drabenstott and Barkema in chapter 3 categorize the breadth of lenders currently serving the agricultural sector (Figure 3.1). It is unclear at the moment which segments of the Harrington and others' agricultural sector that Drabenstott and Barkema's array of lenders will pursue. In other words, if one were to draw lines to connect each lender with a farm type, how would they be drawn? Which farm groups will individual lenders target? Clearly, each lender has market strengths at the moment. But few align with the groupings of farm types depicted. Consequently, significant repositioning, targeting, and product development are likely to occur in the agricultural finance industry.

Development Required of Lenders

Several past practices of lenders will require transformation. First, new credit and underwriting standards must be developed. Measures of financial performance will differ markedly for the diverse types of firms likely to evolve in the agricultural sector. Benchmarks and methods of measuring liquidity, solvency, profitability, and efficiency also will vary across firms. The Farm Financial Standards Task Force developed key measures that unified many disparate methods of determining financial performance of agricultural firms in the late twentieth century. Although useful at the time, they are becoming more difficult to apply as the firms in the sector become more heterogenous, have differing resource combinations, and pursue diverse financial goals. Consider the appropriate financial standards for the following integrated corn firms:

- Firm A contracts with a major food company and raises 1,000 acres of a genetically altered high-starch hybrid for the food industry.
- Firm B markets 500 acres of a specialty popcorn to upscale retail stores on the East Coast.

• Firm C is the "grow agent" of a major pharmaceutical company that has the responsibility of raising 500 acres of seed stock to expand the acreage of a special high-oil corn variety for industrial processing.

Each firm produces a unique and differentiated corn product. Each product has a different value and may require specialized resources for production. It is likely that the capital, profitability, and cash-flow requirements vary as well, necessitating individual financial standards for each firm in the industry. Taken even further, this implies that unique financial performance measures will be required for each type of firm (farm/agribusiness) depicted in the Harrington et al. model in chapter 2 of farm structure in the twenty-first century.

Second, agricultural lenders will have to complete their transition from asset-based to cash-flow-based lending policies and procedures. Integrated firms at the top of the range of farm sizes (Harrington et al.) own few tangible assets. They acquire use of assets primarily through rental and leasing contractual arrangements. On the other end of the Harrington et al. farm size spectrum, part-time farms have assets that are comingled and funded by nonfarm and household activities. In either case, traditional asset-based measures of financial performance provide biased assessments. In the past decade, agricultural lenders have made substantial progress developing and integrating cash-flow-based lending criteria in their credit and underwriting standards. However, credit scoring systems published in popular journals and conversations with major lending institutions confirm heavy reliance on asset-based lending criteria.

Given the integration of farming units envisioned, even systematic methods of evaluation (e.g., credit scoring/screening, uniform underwriting standards, etc.) may be difficult to administer given the heterogeneity of firms, inability to delineate the number of alliances/relationships intertwined with the firm, and the value/proprietary nature of strategic information. An appraisal of an integrated firm's management may become more important than an assessment of physical assets. The subjective nature of such assessments does not fit traditional evaluation methods. Consequently, new loan decision tools and standards need to be developed. In short, lending to these firms appears to require a different set of skills, and perhaps, some refinements to the accounting reports of the businesses. Agriculture can probably learn a great deal from lenders such as General Electric Capital or Rabobank, both of which have developed strong capability in evaluating complex business relationships.

Financial institutions also will need to develop the ability to assess the economic value of business relationships and be prepared to finance their development. To some extent, the principles will be analogous to valuing the intangible asset, goodwill. If, as Boehlje asserts, the true value of large-scale integrated firms stems from the relationships and alliances that are contained therein, these "soft assets" will be the basis on which firms demand financial capital from lending institutions.

Creative new credit instruments will need to be developed and tailored to individual firms. When compared with other midsize commercial businesses, agricultural firms are relatively sophisticated in their use of risk management products, including insurance, options, derivatives, and contracts to mitigate supply, production, and market uncertainties. Similar specialization and containment of financial risk likely will be demanded in credit services. Such products could involve contingencies, conversions, sharing, or be event-specific. Similar to nonfarm businesses, more frequent use of credit lines may be arranged and available, contingent upon specific events occurring (e.g., an investment opportunity in a value-added enterprise). To the extent that agricultural insurance markets evolve in tandem with financial services, indemnities from business interruption policies may serve as security for information-based enterprises, such as crop consulting and fertilizer/pesticide application services, that are comprised of 'soft' intangible assets. Indeed, it is likely that agricultural enterprises will search out and adapt to their use a range of risk management tools being used in other business sectors.

In addition to providing debt financing, financial institutions patronizing large-scale integrated farms will have to develop methods of supplying equity capital as well, including underwriting of stock offerings and other equity products. Given the scale and type of projects likely to be undertaken by these integrated firms, a blend of equity and debt financing probably will be demanded.

Many integrated firms are keenly involved in cutting-edge technology. The largest U.S. technology and pharmaceutical firms currently enjoy relatively favorable access to both debt and equity capital. However, existing markets have difficulty providing financial capital, particularly equity capital, to mid-size and small firms. The monies that are required for new technological research (especially biological research) entail risks that are uncertain, but yield highly profitable returns to innovators that capture the economic rents from commercializing new technology.

Internationalization of the U. S. agricultural finance industry lags that occurring in domestic production agriculture and agribusiness, by a wide margin. At present, firms desiring to import from or export to firms overseas seek letters of credit, currency exchange, country risk assessments, foreign exchange risk management, and other financial services supporting international trade. For these services, firms primarily turn to nonagricultural financial institutions. Relatively few agricultural lenders are able to arrange credit denominated in other currencies, to provide LIBOR-based[8] interest rates, or have experience with credit evaluation of firms that use accounting systems common in other countries. To completely serve the financial needs of emerging farm units, existing agricultural lenders will need to develop these capabilities inhouse, or establish alliances with other financial firms that enable lenders to provide seamless access to such financial services for their agricultural and agribusiness customers.

Box 8.1 A Note on Traditional Agricultural Lenders

The focus of this book has been on future opportunities and challenges for both traditional and nontraditional agricultural lenders. This does not imply, however, the impending demise of rural community banks and the FCS. Each of these traditional lenders have strengths that likely will continue to serve them well as credit markets change.

Community banks appear likely to shrink further in numbers as a result of ongoing mergers and acquisitions. Some of this activity is intended by owners to create stronger, more viable banking organizations. Many of these banks will evolve to serve as the door through which customers can acquire a much broader range of credit and other financial services. These banks probably will prosper in the new environment by doing so. Others will evolve to serving as consumer and small business/small farming lenders. So, despite growing competition across the credit market, there is room for these banks to prosper in niche markets if they focus on their community presence and pursue excellence in what they offer to customers.

The FCS has different opportunities and challenges than commercial banks. If farmers adapt well to distance and computer-based lending, the FCS' delivery system, which is more cost-effective in dealing with larger customers than small or midsized ones, will be an advantage. If not, the lack of community presence will prove to be a hurdle that must be overcome by exceptional service and innovative customer focus. As farms grow larger and more complex, the FCS' capacity to evaluate, fund, and manage large lines of credit will help them seize growth opportunities. However, their current customer base is made up of older farmers with larger farm businesses than is true for commercial banks. Attracting a new generation of customers who have the same loyalty to the FCS that current customers have may prove to be a daunting challenge. Historically, the FCS has offered farmers a superior mortgage credit product, and that business line has accounted for the majority of their outstanding loan volume. Increasingly, the mortgage business line will be challenged by commercial banks and nontraditional lenders, perhaps through the use of a secondary market for farm mortgage credit, such as Farmer Mac. Nevertheless, strong and innovative leadership can chart out a bright future for the FCS.

As public policymakers become more focused on rural revitalization and economic development, both commercial banks and the FCS may be asked to shoulder increased community service responsibilities. For banks, that may mean more attention to community investment and equal-opportunity lending responsibilities. For the FCS, that may mean greater focus on lending to small and low resource farmers, rural housing lending programs, and equal opportunity lending responsibilities.

Finally, financial institutions serving not only agriculture but all sectors of the economy will need to devote resources to developing new delivery mediums, including computer-based banking, use of the Internet for banking transactions, low-documentation loan products, signature loans, and alliances with retailers who administer point-of-sale financing programs. The scale efficiencies of these new technologies imply economic benefits for most agricultural firms and their lenders. However, the greatest beneficiaries may be geographically remote firms that will no longer experience significant distance-related transactions costs— further broadening the set of financial services available to these firms and raising the competitive stakes for community-based lenders and providers of financial services.

Public Policy Implications

The evolution of the agricultural lending industry that is likely to occur in the next decade raises several public policy issues. As history has shown, no single policy will likely meet the needs of all agricultural and agribusiness firms. Instead, agricultural finance policy will likely evolve as a series of carefully targeted programs designed to maintain a competitive industry structure and to address clear instances of market failure. To achieve those objectives, several new policies may come into place to facilitate continued economic integration of the agricultural sector with the rest of the economy. Attention will also be given to the effects of that integration on rural communities and on the environment.

Integrated, Large-Scale Firms and Their Lenders

The ongoing justification for public programs to supply reasonably priced credit to either farm or rural sectors of the country will be challenged on several fronts. First, agricultural firms likely will be neither entirely farm nor rural in their ownership and location. Integrated, large-scale farm units will be comprised of several coordinated input-purchasing, processing, marketing and product distribution entities. Traditional farm production activities may account for only a small portion of a firm's business activity. Consequently, demands for favored public policy by integrated farming units would probably warrant no greater attention than similar demands expressed by other rural commercial firms with linkages to agriculture (e.g., farm equipment or lumber industries). Credit access will be on a commercial basis with methodologies for evaluating firm financial performance and creditworthiness applied uniformly to peer firms in a community, across the nation, and internationally. Integrated farming units probably will not be able to claim rural disadvantage, either, as most will have headquarters in metropolitan areas of significant size.

The need for, and political support of, single-sector financial institutions catering to the industry may also wane. Single-sector financial institutions with special government imprimatur, such as the FCS, increasingly will be asked to demonstrate that they fulfill a public purpose in exchange for that imprimatur. New public-purpose institutions will be increasingly difficult to create, when special public support is required. Conversely, such single-purpose institutions already in place seem likely to be elusive targets for public policymakers desiring to withdraw special public support. Nonetheless, the emergence of viable secondary markets in agricultural and rural business loans may require initial support from the public sector. The Federal Agricultural Mortgage Corporation (Farmer Mac) is a case in point.

The volume of financial services demanded by large-scale integrated farms will exceed the capacity, loan authority, and credit limits of most local financial institutions. Thus, their transactions will no doubt be serviced by the larger financial institutions that already possess well-diversified portfolios and the capacity for evaluating complex business proposals. Likewise, part-time farmers will be most efficiently served by financial institutions emphasizing consumer services with a transactions-based approach rather than a relationship approach to service pricing and delivery. In either case, the demand for specialized agricultural services, as compared with broader-based financial services, appears limited.

By and large, technological advances in the financial services industry have removed most geographic barriers to commerce in either saving or lending activities. Consequently, rural lenders are able to obtain loanable funds at rates that approximate those available to urban firms. Larger lenders appear to have greater flexibility in matching maturities of loanable funds with the maturities of the projects to which they are loaned.

Several industries in the U.S. economy have historically benefited from favored programs of financial assistance that have been developed and administered through financial institutions. An example is the loan guarantees on export credit extended for the purchase of U.S. products, available to exporters through the U.S. Export/Import Bank. Boeing and Caterpillar are two firms that have benefited from its use. Major grain trading firms such as Cargill or Continental Grain have benefited from USDA's Export Credit Guarantee Program, commonly known as the GSM program after the General Sales Manager's office which operates the program, for guaranteeing credit for the purchase of U.S. agricultural commodities. U.S. farmers have benefited as well. The Overseas Private Investment Corporation (OPIC) provides risk insurance for investment by U.S. companies in foreign countries. McDonald's, for example, has made extensive use of such risk management during their rapid international business expansion. Integrated agricultural producers and agribusiness firms also use these tools.

It is unlikely, however, that integrated agricultural firms will be able to establish support for additional financial assistance programs for several reasons. First, federal budget pressures limit the assistance available to new programs. Current benefactors will continue to receive the majority of financial assistance because of their established political contacts and lobbying efforts. Second, even though the agricultural sector is becoming more integrated, the industry has not reached a level of concentration that is of public concern. Although the number of agriculturally related firms listed in the Fortune 500 continues to grow, the sector still is comprised of hundreds of firms that actively compete among themselves. Finally, lenders providing financial services to the largest integrated agricultural firms have demonstrated only limited interest in working cooperatively to establish such programs.

Small and Part-Time Farmers

Small-scale, part-time farms located in rural areas will most likely be operated by individuals with household incomes similar in size to other rural residents in their neighborhoods. It is just a coincidence that their part-time work activities and avocations are agricultural, as opposed to other outdoor sports, hobbies, or rural businesses.

U.S. public policy has long given special attention to small and low-resource agricultural producers. Special loan and loan guarantee programs have been provided through the FSA and its predecessor agency, the FmHA. Despite the special attention in government policy, most benefits of these programs probably have gone to larger-scale agricultural producers and relatively well established extended farming families. In recent years, the declining numbers of small and low-resource farmers have focused even more federal attention on their situation than previously.

Even though public policy has not been successful in maintaining numbers of these farmers, it seems quite likely that government will continue to provide special programs to assist them. The most prominent of these programs will be focused on providing credit to purchase farm real estate and to finance farming operations. Indeed, the ongoing emphasis on reducing government size probably implies that federal government farm loan and loan guarantee programs will narrow their purview and become even more closely focused on the needs of small and low-resource operators.

What will those finance programs look like? Probably a good deal like the current credit programs of the FSA. Limited amounts of direct credit will be available to the most needy of these farmers, with larger amounts of guarantee authority for credit extended by commercial lenders. The emphases of these programs will be twofold. First, small farm operators will be assisted in growing the size of their farm businesses, ideally to the point at which a competitive family income can be earned from the farm and available off farm sources. Second, low-resource farms will be assisted in acquiring the farm business management tools to become competitive agricultural producers. In both cases, most of these farmers will become part-time farmers, earning much of their family income from off-farm sources. That off-farm income may come from custom farming work such as field preparation, planting, pesticide application, or harvesting. It also might include trucking or other activities only tangentially related to farming. Even more likely, the income may come from work in a manufacturing plant or service business that has no relationship to agriculture.

Access to lease financing and equipment rental arrangements will be important to the success of such off-farm employment for the part-time farmer. A wide range of lease firms and equipment manufacturing firms will provide lease financing. Commercial lenders will also provide debt capital, some of which will be loans guaranteed by federal or state governments. Business credit to these producers likely will be patterned after consumer credit. Only limited financial information will be required of the borrower, amounts loaned will be less than $100,000 annually, and interest rates will be 2 to 4 percentage points above those charged commercial-scale farmers, to compensate for the risk associated with limited information on the borrower and borrower's business. Indeed, credit card debt may become a common form of business credit, as has become the case with small-scale, nonfarm businesses. Suppliers and processors, such as Farmland Industries and Continental Grain, also will become more important sources of credit, especially when a farmer is growing livestock or crops on a contract basis for a marketing or processing firm.

Both equity and debt capital may be provided by these firms, as is now true for-contract poultry and hog production. In some cases, loan guarantees will be offered on facility loans from commercial lenders or facility manufacturers by integrators. Multiyear production contracts may also be available for some producers.

Indeed, one might expect much more effort by producers to organize and bargain for improved contract terms and multiyear commitments from integrators and processors. In some cases, the farmer and the contracting firm may develop business relationships that more closely resemble partnerships than arms-length business arrangements.

Finally, off-farm employment by one or more family members will become the typical arrangement for small and limited-resource farmers, if they are to achieve a competitive level of family living. That, of course, implies a much closer linkage between the interests of agricultural producers and those of nonfarm businesses in rural communities. Unfortunately, in most of the country, that relationship has not yet matured. Few farm organizations or commodity groups understand the importance of off-farm employment opportunities in keeping their farm families on the land.

Financing Entry Into Farming

Another public policy issue is the apparent desire of policymakers to maintain ease of access to farm business operation for anyone desiring to do so. This broadly held desire is increasingly unrealistic, however, as farm scale grows ever larger and the capital and management demands of commercial farm operation challenge even the most successful farmers. While we might expect public policymakers to continue federal and state government loan programs, in an effort to achieve equality of access to full-time farm operatorship, the effort will be less successful in the future than in the past.

Nonetheless, a number of new avenues to entering agriculture are increasing the range of opportunities for young people who do not have families well established in agricultural production. Contract production, most typically of livestock such as poultry and swine, represents a growing opportunity for new entrants into farming, and can often be achieved with limited resource investment by the new farmer. In many instances, contracting firms provide partial financing of required facilities. In other cases, contractual arrangements along with limited owner equity can enable the new entrant to obtain sufficient debt capital to enter the business.

Here again, lease financing of facilities can be an attractive alternative to facility ownership. Facility and equipment manufacturers sometimes provide such lease financing. It is likely that public support will increase for a broad-based financial secondary market, capable of securitizing facility and major equipment loans or leases. Borrowers seem likely to seek more of that type of financial service, and lenders will seek improved means to better manage the risk inherent in such credit and lease financing.

A wide array of lenders provide both debt capital and lease financing arrangements to firms; although neither federal nor state loan programs have, as yet, been developed to support such business development.

Environmental Issues for Lenders

Lenders have become increasingly sensitive to environmental pollution problems of business firms to which they extend debt capital. Typically, lenders now avoid knowingly including borrower assets in collateral that secures a loan when those assets embody significant environmental problems. Lenders also remain exceedingly cautious about taking possession of property in foreclosure when that property is known to have significant environmental problems.

Courts have, in some cases, held lenders responsible for environmental cleanup when they have become the owners of record as a result of foreclosure, even though the lenders had no role in creating the pollution problem. Crafting legislation and regulations in ways that do not preclude a prudent lender from providing debt capital for such enterprises as confinement livestock production and specialized crop production requiring substantial pesticide use will be important.

The increasing use of large-scale confinement facilities for poultry and swine production is creating new pollution control issues for lenders. Animal waste storage can be subject to leakage into groundwater, overflow from storage facilities can pollute streams, and odor from facilities often prove both offensive to neighbors and litigious to the producers. Lenders now routinely perform environmental scans on firms to which they will be lending, if there is risk of environmental cleanup problems on firm assets that provide collateral for a loan. Full environmental audits are required where there is concern that significant environmental hazards may exist on property that serves as loan collateral, or that may be acquired by a lender in foreclosure. Lenders understandably weigh the value of such collateral in reducing loss on an unpaid loan against the potential liability for environmental cleanup of the property.

Lenders also seek to understand environmental risks experienced in the conduct of a customer's business, since the cost of environmental cleanup, or the legal liabilities that can result from business mismanagement that causes environmental problems, can financially ruin the business. Producing and marketing unsafe products, such as meat products containing dangerous bacteria, can result in loss of customers and government fines that destabilize or destroy a lender's business customers. Hence, lenders must develop a keen understanding of the environmental and food safety risks inherent in the business practices of their customers, and must develop an acceptable level of comfort that the customer can manage such risk successfully.

As public awareness of environmental issues increases, agricultural firms likely will be held to higher standards of environmental protection and safety of products. Legislative and regulatory actions to protect consumers and the environment will be closely monitored by lenders. Increased use of zoning ordinances, to authorize or to preclude certain agricultural production practices, and licensing, to ensure compliance with environmental standards, seem likely as part of public management of agricultural production/processing where pollution may be a byproduct. New legislation and case law will more clearly define the responsibility of lenders for the environmentally damaging actions of their customers.

Ownership and Control of Industrialized Agriculture

In many states, laws restricting the form of ownership of agricultural real estate or farming remain in effect. Laws restricting foreign ownership and corporate ownership of farmland and farm businesses were enacted ostensibly to protect smaller-scale family farm ownership and control of agriculture. Because of changes in technology applied to agriculture and agribusiness, increases in scale of firms involved, and international ownership of the firms, the restrictive federal and state laws increasingly are at odds with successful business development and operation. In a number of states where anti-corporate farming laws remain from earlier decades, these laws may now be impeding farmers' access to equity capital and to off-farm equity partners to share the risk associated with complex and costly production systems.

As the cost of entry into certain types of agricultural production, such as large-scale confinement swine production, continues to rise for individual farmers, they often are unable to secure sufficient equity and, sometimes, debt capital to enter these businesses. Frequently, farmers are reluctant to pledge all their assets in order to obtain sufficient credit for large-scale livestock production systems, and would welcome an equity partner to spread the risk associated with the venture.

As the scale of production continues to grow, both for livestock and crop production, it seems less likely that the traditional family farmer will fully own and control the emerging farm production systems. Reaching a broadly acceptable venue in which both ownership and control of farm businesses are increasingly likely to be shared by traditional family farmers with off-farm investors and input suppliers/processors/marketers promises to be a difficult and contentious process. Yet, it is a necessary evolution if farmers are to fully participate in the increasingly integrated, interconnected, and sophisticated agricultural production systems that are emerging across much of the country. How the issues of control and ownership are resolved is also of great import to agricultural lenders as they seek to position themselves as providers of debt capital and other financial services to the agricultural sector.

Rural Development Financing

Most established rural businesses are creditworthy and possess adequate access to reasonably priced capital and financial services. However, rural financial markets do have shortcomings. Rural firms that are undercapitalized, becoming established, expanding, minority-owned, or that operate in niche industries are generally perceived to have greater difficulty obtaining financial capital, compared with peer urban firms. Moreover, these difficulties are exacerbated during periods of economic recession. Even during periods of economic expansion, some community banks face greater loan demand than their deposit base supports. And, there often is a mismatch between the maturity of loanable funds a community bank has available to support a loan and the maturity of economic development loans sought by rural business borrowers. Economic development loans usually are repaid over a much longer period than the maturity of banks' loanable funds, which usually is 2 years or less. Rural businesses have fewer capital sources and are more dependent on community banks. Sources of equity capital are even more difficult to access.

Several remedies have been proposed to increase the supply of capital available to rural firms experiencing credit rationing. First, existing loan programs and the charters of rural financial institutions could be modified to expand their loan-making authority to disadvantaged rural businesses. This would require rural lenders to make more loans to businesses that have difficulty in obtaining necessary financial capital and services. Of course, the origination of riskier loans could jeopardize the financial strength of these rural lenders, unless other forms of assistance were forthcoming to offset their potential loan losses and capital erosion.

Another popular suggestion is to expand the development of rural secondary financial markets. Qualifying loans could be used as collateral to support securities sold into a secondary market. An expanded role for Farmer Mac in securitizing rural economic development loans has been proposed by some public policymakers.

Alternatively, a community- or state-based revolving loan fund could be established with federal support. Loans made from these funds could then be sold into the secondary market with the proceeds being recycled back into the loan fund. However, several important challenges face the development of this structure. The market will no doubt be small-scale. Comprehensive financial performance information necessary to underwrite these loans currently is unavailable. Finally, an industry network of poolers and service agents is not yet in place.

Areas of Increasing Public Policy Concern

Two areas of concern and need for public policy remain. First, as the agricultural sector progresses through rapid change in the next decade toward the structure envisioned by Harrington and others in chapter 2 and Drabenstott and Barkema in chapter 3, public policies may need to be developed to assist this transition. What public policies would be useful and cost-effective? How long should the public sector continue its involvement and support of specialized agricultural finance institutions? How does the public sector minimize or otherwise manage the abrupt dissolution of financial commitments to existing firms? Or, should it be concerned? Is the private sector better equipped to address these transition issues? These questions all will require answers as policymakers consider whether and how transition assistance should be provided to persons in the agricultural sector.

The second area of concern involves possible contagion to the agricultural sector and the nation's food and fiber sector from the failure of large-scale integrated agricultural firms. Bankruptcy of traditional farm units had, at most, spillover effects on the local economic base of the community patronized by the firm. Spillover from the failure of a large-scale integrated agricultural firm would no doubt be more broadly experienced. The breadth of disruption could span consumers, related industry, and financial service sectors and range from regional to national and perhaps international dimensions. Should public and private policymakers be concerned about disruptions caused by failure of large integrated firms? If so, what should be done? Currently, there is no consensus as to what should be done, or indeed, if anything should be done.

Conclusion

Current trends affecting the structure, scale, and organization of farms and their relationships with suppliers and processors/marketers suggest unprecedented changes for agricultural lenders. Federal farm credit programs will primarily target the least creditworthy operations, as they did prior to 1970. The federal role in agricultural credit markets remains important, with approximately 11 percent of farm debt provided either through direct loans or indirectly through loan guarantees of the FSA and Small Business Administration (SBA). The federal government also has a presence in farm lending through the activities of the FCS and Farmer Mac as GSEs. FSA and SBA loan market share likely will decline in the new century, in the face of federal budget limitations and because of the transition of farm structure toward larger and more industrialized units, but the future farm lending market share trends of the FCS and Farmer Mac are less clear.

All farm lenders will face greater diversity among borrowers. Those borrowers involved in integrated agricultural production systems will require larger lines of credit for businesses that are more complex and more difficult to evaluate vis-a-vis financial performance and creditworthiness. A broader range of financial services will no doubt be required. Yet, small-scale, part-time farm businesses may also grow in number. Finally, both types of farm organization will exist alongside traditional family farms. Business growth opportunities for lenders are more likely to be found among small-scale, part-time farms and large integrated farms, however. The ongoing shift from asset-based lending to cash-flow-based lending will continue to challenge lenders. Lenders will experience more competition from among their traditional competitors and, especially, from nontraditional lenders and foreign lenders doing business in the United States. All told, the environment for agricultural lending will be challenging, but profitable, for those firms that are able to add new value for their customers.

Notes

1. Financial intermediation's role is to conduct the transfer and rationing functions in financial markets. The time, place, form, and other characteristics of funds provided by suppliers is transferred to make them available for users. Suppliers and users want different yields, liquidity, maturity, timing, and risk characteristics. Financial intermediaries actually perform several different kinds of intermediation. These include: (1) denomination intermediation, (2) default-risk intermediation, (3) maturity intermediation, (4) information intermediation, (5) risk pooling, and (6) economies of scale (Rose and Fraser 1985).

2. In 1933, under the Federal Emergency Relief Act (P.L. 73-15) and the Farm Credit Act (P.L. 73-75), the entire system was reorganized and the privately owned intermediaries, the joint-stock banks, liquidated because of a number of factors with the primary one being the number of constraints placed on their operations that caused them to fail (O'Hara 1983).

3. The increase in the growth rate of total farm business debt is instructive regarding events of 1960-85. Total nominal farm business debt grew 15.7 percent 1940-49, 96.2 percent 1950-59, 106.8 percent 1960-69, and 210.9 percent 1970-79. It expanded 297.5 percent from 1970 to the 1984 peak, before falling 28.9 percent from 1984 to 1989. It increased 17.7 percent 1989-97.

4. Financial stress can have a variety of meanings, but it generally is regarded as when a farm household does not have sufficient cash to meet the cash expenses of the farm operation, family living, and scheduled debt service.

5. Data are unavailable for life insurance companies, but their farm mortgage loan charge-offs were substantial. Hanson et al. (1991) estimated their farm loan charge-offs at $859 million for 1984-89.

6. The quest for increased public sources of loan funds for U.S. agricultural lenders and farmers historically has extended beyond federal government credit programs and sources. For example, in 1993-94, according to surveys conducted by USDA's Economic Research Service, 32 states operated 81 agricultural credit programs with preferential terms for farmers and ranchers (Wallace et al. 1994). For a program to be included, loan proceeds had to be used to purchase land, machinery, or other operating inputs used in the agricultural production enterprises. The outstanding loan balance for all such state programs in 1993-94 was $1.8 billion.

7. Collender and Koenig observe in chapter 6 that the FCS focuses its lending "on larger commercial farming operations, perhaps giving these producers an additional competitive advantage relative to small and medium-sized operations" (p. 144). The FCS' Farm Credit Banks and Associations on June 30, 1997, had 590,276 loans (this excludes Bank of Cooperative loans). This compares with 2.1 million farms in 1996, of which 537,480 had annual gross farm sales of $50,000 or above (USDA, 1997b).

8. LIBOR is the London Inter-Bank Offered Rate. It is the interest rate on dollar deposits loaned between first class banks in London. Its principal use is as the base interest rate on which prices of Eurodollar and other Eurocurrency loans are calculated. It is a prime bankers' rate and is often used in international banking as a basic rate in referring to an interest rate being negotiated. The International Monetary Fund uses it as a benchmark for calculating the interest rate on most of its lending. These loans specify an agreed spread above the LIBOR 3- or 6-month rate. There is no set procedure or set time for changing LIBOR.

References

Barry, Peter J. 1995. *The Effects of Credit Policies on U.S. Agriculture*. Washington, D.C.: The American Enterprise Institute Press.

Barry, Peter J., and Michael D. Boehlje. 1986. "Farm Financial Policy," in Dean W. Hughes, Stephen C. Gabriel, Peter J. Barry, and Michael D. Boehlje (eds.), *Financing The Agricultural Sector: Future Challenges and Policy Alternatives*, Boulder, CO: Westview Press, pp. 129-174.

Brake, John R. 1974. "A Perspective on Federal Involvement in Agricultural Credit Programs," *South Dakota Law Review*, (Summer) 19: 567-602.

Cochrane, Willard W. 1979. *The Development of American Agriculture: A Historical Analysis*. Minneapolis, MN: University of Minnesota Press.

Collender, Robert N. 1992. "Changes in Farm Credit System Structure," in U.S. Dept. Agr., Econ. Res. Serv., *Agricultural Income and Finance: Situation and Outlook Report.* AFO-44 (Feb.), pp. 37-44.

Collender, Robert N. 1991. "Have Mergers Improved the Financial Performance of Farm Credit Banks?" in U.S. Dept. Agr., Econ. Res. Serv., *Agricultural Income and Finance: Situation and Outlook Report.* AFO-40 (Feb.), pp. 35-40.

Collender, Robert N., and Audrae Erickson. 1996. *Farm Credit System Safety and Soundness.* AIB-722. U.S. Dept. Agr., Econ. Res. Serv. (Jan.).

Dyson, Lowell K. 1988. "History of Federal Drought Relief Programs." Staff Rept. No. AGES880914. U.S. Dept. Agr., Econ. Res. Serv., Oct.

Harrington, David H., Leslie Whitener, Ray D. Bollman, David Freshwater, and Philip Ehrensaft (eds.). 1995. "Farms, Farm Families, and Farming Communities," *Canadian Journal of Agricultural Economics,* 1995 Special Issue, pp. 1-253.

Hiemstra, Stephen W., Steven R. Koenig, and David Freshwater. 1988. *Prospects for a Secondary Market for Farm Mortgages.* AER-603. U.S. Dept. Agr., Econ. Res. Serv., Dec.

Hanson, Gregory D., G. Hossein Parandvash, and James Ryan. 1991. *Loan Repayment Problems of Farmers in the Mid-1980's.* AER-649. U.S. Dept. Agr., Econ. Res. Serv. (Sept.).

Hoag, W. Gifford. 1976. *The Farm Credit System: A History of Financial Self Help.* Danville, IL: The Interstate Printers and Publishers, Inc.

Holmes, Wendell, Thomas Carlin, and Margaret Butler, (1991). Article in U.S. Dept. Agr., Econ. Res. Serv., *Agricultural Income and Finance: Situation and Outlook Report.* AFO-43 (Dec.), pp. 11-14.

McKie, James W. 1963. "Credit Gaps and Federal Credit Programs," *Federal Credit Programs.* Research Study Three. Commission on Money and Credit. Englewood Cliffs, NJ: Prentice-Hall, Inc., pp. 317-353.

Moss, LeeAnn McEdwards, Peter J. Barry, and Paul N. Ellinger. 1997. "The Competitive Environment for Agricultural Banks in the US," *Agribusiness* 13 (4): 431-444.

Lee, Warren F., and George D. Irwin. 1996. "Restructuring the Farm Credit System: A Progress Report," *Agricultural Finance Review,* Vol. 56, Ithaca, NY: Department of Agricultural., Resource, and Managerial Economics, Cornell University, pp. 1-21.

Lins, David A. 1979. "Credit Availability Effects On the Structure of Farming," in *Structural Issues Of American Agriculture.* U.S. Dept. Agr., Econ. Res. Serv., Nov., pp. 134-142.

Nourse, Edwin G. 1916. *Agricultural Economics.* Chicago: University of Chicago Press.

O'Hara, Maureen. 1983. "Tax-Exempt Financing: Some Lessons from History, " *Journal of Money, Credit and Banking* (Nov.) 15 (4): 425-441.

Peoples, Kenneth L., David Freshwater, Gregory D. Hanson, Paul T. Prentice, and Eric P. Thor. 1992. *Anatomy of an American Agricultural Credit Crisis.* Lanham, MD: Rowman and Littlefield Publishers, Inc.

Rose, Peter S., and Donald R. Fraser. 1985. *Financial Institutions.* Second Edition, Plano, TX: Business Publications, Inc.

Sommer, Judith E., David E. Banker, Robert C. Green, Judith Z. Kalbacher, R. Neal Peterson, and Theresa Y. Sun. 1997. *Structural and Financial Characteristics of U.S. Farms, 1994: 19th Annual Family Farm Report to the Congress.* AIB-735. U.S. Dept. Agr., Econ. Res. Serv., July.

Sparks, Earl Sylvester. 1932. *History and Theory of Agricultural Credit in the United States.* New York: Thomas Y. Crowley Company.

Spong, Kenneth. 1990. *Banking Regulation: Its Purposes, Implementation, and Effects.* Third Ed. Division of Bank Supervision and Structure, Federal Reserve Bank of Kansas City.

Stam, Jerome M., Steven R. Koenig, Susan E. Bentley, and H. Frederick Gale, Jr. 1991. *Farm Financial Stress, Farm Exits, and Public Sector Assistance to the Farm Sector in the 1980's.* AER-645. U.S. Dept. Agr., Econ. Res. Serv., April.

Stam, Jerome M., Steven R. Koenig, and George B. Wallace. 1995. *Life Insurance Company Mortgage Lending to U.S. Agriculture: Challenges and Opportunities.* AER-725. U.S. Dept. Agr., Econ. Res. Serv., Dec.

U.S. Country Life Commission. 1909. *Report of the Country Life Commission: Special Message From the President of the United States Transmitting the Report of the Country Life Commission.* 16th Congress, Second Session. Washington, D.C.: U.S. Government Printing Office, Feb.

U.S. Department of Agriculture, Economic Research Service. 1997a. *Agricultural Income and Finance: Situation and Outlook Report.* AIS-65 (June).

U.S. Department of Agriculture, Economic Research Service. 1997b. *Agricultural Income and Finance: Situation and Outlook Report.* AIS-66 (Sept.).

U.S. Department of Agriculture, Economic Research Service. 1997c. *Credit in Rural America.* AER-749, April.

U.S. Department of Agriculture, Economic Research Service. 1984. *Federal Credit Programs for Agriculture.* AIB-483 Nov.

U.S. Department of Agriculture, Farmers Home Administration. 1989. *A Brief History of Farmers Home Administration.* Feb.

U.S. Department of Agriculture, Farmers Home Administration. 1991. *Farmers Home Administration: Report for Fiscal 1990.* Feb.

Wallace, George B., Audrae Erickson, and James J. Mikesell. 1994. "Handbook of State-Sponsored Agricultural Credit Programs." ERS Staff Paper No. 9426. U.S. Dept. Agr., Econ. Res. Serv., Nov.

About the Book and Editors

This volume is concerned with the paradigm shifts occurring in U.S. agriculture and its related financial services sector. The U.S. agricultural sector is undergoing rapid change with large segments commonly described as industrialized. Often observers focus on the technological and structural changes that the sector is undergoing and ignore other critical change factors, such as financial services variables. U.S. agriculture has long been a capital-intensive sector and it is becoming more so. Financing this sector requires a complex set of financial arrangements that involve both debt and equity capital. Total U.S. farm sector business debt totaled $162 billion at the end of 1997, and the financial flows involving the agricultural sector are becoming increasingly intricate.

Industrialization in agriculture is commonly defined as the increasing consolidation of farms accompanied by increasing vertical coordination among the stages in the production system through contracting and integration, all driven by changes in consumer demand, production technology, and international demand and competition. The concentration of the U.S. agricultural complex—with fewer entities controlling larger segments of the market—is seen in a wide range of agricultural activities, including the poultry, beef, swine, dairy, sugar, feed, seed, and fertilizer markets. In 1996, some 336, 400 farms with incomes exceeding $100,000 per year accounted for 16.3 percent of all farms and 81.5 percent of all cash receipts from farm marketings. The largest farm firms look increasingly like firms in other sectors of the American economy, both in terms of the size of business units and in their organization as reflected in contractual and financial relationships. Vertical integration is important in dairy, seed, fruit and vegetable, turkey, egg, and broiler production. About 20 percent of beef and pork production takes place under contract. These changes affect credit use and how farm sector financing is conducted.

A small farm or part-time farm subsector exists. About 1.5 million farms could be considered small or noncommercial, with annual gross cash sales of less than $50,000. Small farms fall roughly into four groups by the age of the primary operator and the amounts and sources of off-farm income: part-time, country gentleman, retired, and subsistence. These groups differ in their demand for farm credit, but overall they are not a major factor in total credit use.

Lenders operating in rural areas face numerous challenges in financing agriculture into the twenty-first century as the agricultural sector goes through a rapid transformation to a new paradigm. Lenders serving U.S. agriculture continue to undergo rapid change. The commercial bank sector continues to consolidate and modernize. There is a continuing impact from deregulation and new technology. The Farm Credit System has undergone a major consolidation during the last decade while recovering from the farm financial crisis of the early to mid-1980s. The number of associations and banks has declined dramatically. The Farm Service Agency (FSA—formerly the Farmers Home Administration) has assumed more of its traditional role as a lender of last resort. FSA's total direct farm loans outstanding declined dramatically after 1988. The emphasis continues to be on providing loan guarantees.

The life insurance company farm loan portfolio has been maintained at about $10 billion, or 11 to 12 percent of farm real estate loans. But there has been a major effort to concentrate on large-scale agriculture during the last decade. The Pacific region, Florida, and Texas together now account for about 55 percent of the total dollar volume of outstanding life insurance farm mortgages. The "individuals and others" loan category has accounted for a growing share of farm debt. There is much interest in the growth, dynamism, and innovation in the noninstitutional lender or trade credit component of farm lending.

The first major section of this volume investigates forces inducing change. Glenn D. Pederson, Jeffrey T. Stensland, and Martin L. Fischer discuss macroeconomic factors and international linkages affecting the financing of agriculture in a world economy. David H. Harrington, Robert A. Hoppe, R. Neal Peterson, David Banker, and H. Frederick Gale, Jr. look at the major changes in U.S. farm sector structure. The second section of the book focuses on future directions for agricultural finance. Mark Drabenstott and Alan Barkema investigate new and changing rules of the game. Michael D. Boehlje provides insights into the emerging agricultural lending system. Emerging strategies for traditional lenders are outlined by Allen M. Featherstone, Michael D. Boehlje, and Joseph O. Arata. The third section of the book addresses some of the questions regarding future agricultural lenders and the range of their activities. Robert N. Collender and Steven R. Koenig discuss the emerging role of federal agricultural credit programs. Bruce J. Sherrick looks at the emerging nontraditional agricultural lenders. In the final major section, Cole R. Gustafson, Marvin Duncan, and Jerome M. Stam outline public and private policy implications of the emerging U.S. agricultural sector and its financing.

Marvin Duncan is professor of agricultural economics at North Dakota State University (NDSU). His work there focuses on rural financial markets, agricultural credit issues and institutions, rural economic development, and rural/agricultural policy. Dr. Duncan has been chairperson in the Department of Agricultural Economics at NDSU, as well. Prior to joining the NDSU faculty, he was a member of the board of the Farm Credit Administration (FCA), a presidentially appointed position. During his service on the FCA Board, spanning the tumultuous years of the mid- to late 1980s, he was active in working with Congress as it developed legislative solutions to the financial problems of the U.S. Farm Credit System and in providing policy leadership in crafting the regulatory framework to implement that legislation. Duncan previously had been appointed Senior Deputy Governor at the FCA. Duncan had been a vice president and economist at the Federal Reserve Bank of Kansas City prior to joining the FCA where, as a member of the Bank's Research Department, he led its research and analysis of the real sector of the Tenth Federal Reserve District economy. Duncan is widely published on issues within his research focus and is a sought-after commentator on issues of rural financial markets, agricultural credit, and agricultural and rural policy. Duncan has been a visiting scholar at the Economic Research Service of the U.S. Department of Agriculture and has been on a USDA advisory committee. Duncan holds B.S. and M.S. degrees from NDSU and a Ph.D. in agricultural economics from Iowa State University.

Jerome Stam is Leader of the Finance Team in the Economic Research Service (ERS) of the U.S. Department of Agriculture, located in Washington, DC. His work involves conducting and managing an array of research, staff, and situation-outlook work dealing with agricultural credit policy, lending institutions, and financial innovations. He began his professional career with the field staff of ERS at St. Paul, Minnesota. During this five-year period, he also was an adjunct assistant professor and later associate professor in the Department of Agricultural and Applied Economics at the University of Minnesota. He has researched regional economics, economic development, public finance, and agricultural finance issues during his career. He has served as the secretary of the National Agricultural Credit Committee since 1984. Dr. Stam holds a B.S. in agriculture from the University of Nebraska, an M.S. and Ph.D. in agricultural economics from Michigan State University, where he was a National Defense Education Act Fellow, and a Master in Public Administration (MPA) degree from Harvard University. He has authored or coauthored numerous publications, articles, and papers.

About the Contributors

Joseph O. Arata is assistant professor of agricultural economics in the Department of Agricultural Economics at Kansas State University in Manhattan, Kansas.

David E. Banker is an Agricultural Economist, Resource Economics Division, Economic Research Service, U.S. Department of Agriculture, Washington, D.C.

Alan D. Barkema is professor and head of the Department of Agricultural Economics at Oklahoma State University in Stillwater, Oklahoma.

Michael D. Boehlje is professor of agricultural economics in the Department of Agricultural Economics at Purdue University in West Lafayette, Indiana.

Robert N. Collender is Senior Financial Economist, Food and Rural Economy Division, Economic Research Service, U.S. Department of Agriculture, Washington, D.C.

Mark Drabenstott is a vice president and economist with the Federal Reserve Bank of Kansas City.

Marvin Duncan is professor of agricultural economics in the Department of Agricultural Economics at North Dakota State University in Fargo, North Dakota.

Allen M. Featherstone is professor of agricultural economics in the Department of Agricultural Economics at Kansas State University in Manhattan, Kansas.

Martin L. Fischer is director of asset/liability risk management at AgriBank Farm Credit Bank in St. Paul, Minnesota.

H. Frederick Gale, Jr. is an Agricultural Economist, Food and Rural Economics Division, Economic Research Service, U.S. Department of Agriculture, Washington, D.C.

Cole R. Gustafson is professor and head of the Department of Agricultural Economics at North Dakota State University in Fargo, North Dakota.

David H. Harrington is a Senior Economist, Resource Economics Division, Economic Research Service, U.S. Department of Agriculture, Washington, D.C.

Robert A. Hoppe is an Agricultural Economist, Resource Economics Division, Economic Research Service, U.S. Department of Agriculture, Washington, D.C.

Steven R. Koenig is Financial Economist, Food and Rural Economy Division, Economic Research Service, U.S. Department of Agriculture, Washington, D.C.

A. Gene Nelson is professor and head of the Department of Agricultural Economics at Texas A & M University in College Station, Texas.

Glenn D. Pederson is professor in the Department of Applied Economics at the University of Minnesota in St. Paul, Minnesota.

R. Neal Peterson was an Agricultural Economist, Resource Economics Division, Economic Research Service, U.S. Department of Agriculture,

Washington, D.C. at the time of writing. He resigned in December 1997 to head a private business venture.

Bruce J. Sherrick is associate professor in the Department of Agricultural Economics at the University of Illinois in Urbana, Illinois.

Jerome M. Stam is leader of the finance team, Food and Rural Economics Division, Economic Research Service, U.S. Department of Agriculture, Washington, D.C.

Jeffrey T. Stensland is a graduate student in the Department of Applied Economics at the University of Minnesota in St. Paul, Minnesota.

Printed and bound by CPI Group (UK) Ltd, Croydon, CR0 4YY

28/10/2024

01780298-0003